# Spon's Estimating Costs Gu
# Plumbing and Heating

**Other books by Bryan Spain**

**also available from Spon Press**

---

*Spon's Estimating Costs Guide to Electrical Works*
(2004 edition)
Pb: 0-415-31853-X

*Spon's Estimating Costs Guide to Minor Works, Alterations and Repairs to Fire, Flood, Gale and Theft Damage*
(2004 edition)
Pb: 0-415-31854-8

*Spon's House Improvement Price Book: house extensions, storm damage work, alterations, loft conversions and insulation* (2003 edition)
Pb: 0-415-30938-7

*Spon's First Stage Estimating Price Book* (2000 edition)
Pb: 0-415-23436-0

*Spon's Construction Resource Handbook* (1998 edition)
Hb: 0-419-23680-5

---

# Spon's Estimating Costs Guide to Plumbing and Heating

## Unit rates and total project costs

**Second edition**

**Bryan Spain**

Spon's Contractors' Handbooks

Spon Press
Taylor & Francis Group
LONDON AND NEW YORK

First published 1999 by Spon Press
Second edition 2004
2 Park Square, Milton Park, Abingdon, Oxon, OX14 4RN

Simultaneously published in the USA and Canada
by Spon Press
270 Madison Ave, New York, NY 10016

Reprinted 2005

*Spon Press is an imprint of the Taylor & Francis Group*



*Publisher's Note*
This book has been prepared from camera-ready copy supplied by the author.

Printed and bound in Great Britain by TJ International Ltd, Padstow, Cornwall.



*British Library Cataloguing in Publication Data*
A catalogue record for this book is available from the British Library

*Library of Congress Cataloging in Publication Data*
A catalog record for this book has been requested

ISBN 0-415-31855-6

# Contents


| | | |
|---|---|---|
| Preface | | xiii |
| Introduction | | xv |
| **Part One: Unit rates** | | 1 |
| H71 | Lead sheet coverings | 3 |
| N13 | Sanitary appliances/fittings | 9 |
| P31 | Holes, chases, covers and supports for services | 16 |
| R10 | Rainwater pipework/gutters | 19 |
| R11 | Drainage above ground | 26 |
| R12 | Drainage below ground | 33 |
| S10 | Hot and cold water | 40 |
| T10 | Gas/oil-fired boilers | 66 |
| T30 | Radiators | 70 |
| Alterations and repairs | | 73 |
| **Part Two: Project costs** | | 83 |
| **PVC-U rainwater goods** | | 85 |
| One-storey gable end | | |
| | terraced, overall size 7 × 6m | 85 |
| | semi-detached, overall 8 × 7m | 85 |

**One-storey gable end (cont'd)**

detached, overall 9 × 8m 86

Two-storey gable end

terraced, overall size 7 × 7m 87

semi-detached, overall 8 × 7m 87

detached, overall 9 × 8m 88

Three-storey gable end

terraced, overall size 8 × 8m 89

semi-detached, overall 9 × 8m 89

detached, overall 10 × 9m 90

One-storey hipped end

terraced, overall size 7 × 6m 91

semi-detached, overall 8 × 7m 91

detached, overall 9 × 7m 92

Two-storey hipped end

terraced, overall size 7 × 7m 93

semi-detached, overall 8 × 7m 93

detached, overall 9 × 8m 94

Three-storey hipped end

terraced, overall size 8 × 8m 95

semi-detached, overall 9 × 8m 95

detached, overall 10 × 9m 96

**Cast iron rainwater goods**

One-storey gable end

terraced, overall size 7 × 6m 97

semi-detached, overall 8 × 7m 97

detached, overall 9 × 8m 98

Two-storey gable end

terraced, overall size 7 × 7m 99

semi-detached, overall 8 × 7m 99

detached, overall 9 × 8m 100

Three-storey gable end

terraced, overall size 8 × 8m 101

semi-detached, overall 9 × 8m 101

detached, overall 10 × 9m 102

One-storey hipped end

terraced, overall size 7 × 6m 103

semi-detached, overall 8 × 7m 103

detached, overall 9 × 7m 104

Two-storey hipped end

terraced, overall size 7 × 7m 105

semi-detached, overall 8 × 7m 105

detached, overall 9 × 8m 106

Three-storey hipped end

terraced, overall size 8 × 8m 107

semi-detached, overall 9 × 8m 107

detached, overall 10 × 9m 108

**Aluminium rainwater goods**

One-storey gable end

terraced, overall size 7 × 6m 109

semi-detached, overall 8 × 7m 109

detached, overall 9 × 7m 110

Two-storey gable end

terraced, overall size 7 × 7m 111

semi-detached, overall 8 × 7m 111

detached, overall 9 × 8m 112

Three-storey gable end

terraced, overall size 8 × 8m 113

semi-detached, overall 9 × 8m 113

detached, overall 10 × 9m 114

One-storey hipped end

terraced, overall size 7 × 6m 115

semi-detached, overall 8 × 7m 115

detached, overall 9 × 7m 116

Two-storey hipped end

terraced, overall size 7 × 7m 117

semi-detached, overall 8 × 7m 117

detached, overall 9 × 8m 118

Three-storey hipped end

terraced, overall size 8 × 8m 119

semi-detached, overall 9 × 8m 119

detached, overall 10 × 9m 120

**Summary of rainwater goods project costs**

PVC-U 121

cast iron 122

aluminium 123

## Bathrooms

Bath with hand shower, WC and lavatory basin 125

Bath with hand shower, WC, lavatory basin and bidet 127

Bath with hand shower, lavatory basin and bidet 129

Bath with hand shower and lavatory basin 131

Bath with hand shower, WC, lavatory basin and shower cubicle 133

Bath with hand shower, WC, lavatory basin, shower cubicle and bidet 135

Bath with hand shower, lavatory basin, shower cubicle and bidet 138

Bath with hand shower, lavatory basin and shower cubicle 140

Bath, WC and lavatory basin 142

Bath, WC, lavatory basin and bidet 144

Bath, lavatory basin and bidet 147

Bath and lavatory basin 149

Bath, WC, lavatory basin and shower cubicle 151

Bath, WC, lavatory basin, shower cubicle and bidet 153

Bath, lavatory basin, shower cubicle and bidet 156

Bath, lavatory basin and shower cubicle 158

**Summary of bathroom project costs** 161

## External waste systems

PVC-U wastes with solvent-welded joints

- one-storey house 163
- two-storey house 164
- three-storey house 165

PVC-U wastes with push-fit joints

- one-storey house 166
- two-storey house 167
- three-storey house 168

PVC-U soil pipes with solvent-welded joints

one-storey house 169

two-storey house 169

three-storey house 170

Cast iron soil pipes with flexible joints

one-storey house 171

two-storey house 171

three-storey house 172

**Summary of external waste project costs** 173

## Central heating systems

One-storey house overall size 7 × 6m 175

One-storey house overall size 8 × 7m 176

One-storey house overall size 9 × 7m 178

Two-storey house overall size 7 × 7m 180

Two-storey house overall size 8 × 7m 182

Two-storey house overall size 9 × 8m 184

Three-storey house overall size 8 × 8m 186

Three-storey house overall size 9 × 8m 188

Three-storey house overall size 10 × 9m 190

**Summary of central heating projects** 193

## Hot and cold water supply systems

One-storey house 195

Two-storey house 196

Three-storey house 198

**Summary of hot and cold water projects** 200

**Part Three: Business matters** 201

Starting a business 203

Running a business 219

Taxation 225

**Part Four: General data** 235

General construction data 237

**Index** 243

# Preface

A small to medium-sized plumbing business competing in today's commercial world needs all the help it can get. The pressure of obtaining work, carrying it out efficiently and ensuring that full payment is made on time is the responsibility of the firm's principal. Any source of information that can help in saving time must be welcome and this book aims to fulfil this need.

Plumbers and heating engineers have had access to price books for many years but they were always obliged to prepare their own quantities and extract the relevant priced items from the books in order to prepare an estimate. This procedure is time-consuming and it is particularly galling when the contractor suspects that the enquiry is not genuine but merely a ruse to check a previously accepted estimate.

This book provides detailed costs of projects in addition to setting out analytical unit rates. It is accepted that it is unlikely that the projects selected will be identical in size to those received in actual enquiries but the range of projects covered should enable a contractor to provide a quick and accurate response to an enquiry by making slight adjustments to the examples provided.

A detailed quotation could then be prepared if the client showed interest in the estimate. This should make an appreciable saving in the one commodity that all contractors are short of…time!

I have received a great deal of support in the research necessary for this type of book and I am grateful to those individuals and firms who have provided the cost data and other information. In particular, I am indebted to Mark Loughrey of Youds, Ellison & Co., Chartered Accountants of Hoylake (tel: 0151-632 3298 or www.yesl.uk.com) who are specialists in advising small businesses. Their research for the information in the business section is based on tax legislation in force in December 2003.

Although every care has been taken in the preparation of the book, neither the publishers nor I can accept any responsibility for the use of the information provided by any firm or individual. Finally, I would welcome any constructive criticism of the book's contents and suggestions that could be incorporated into future editions.

Bryan Spain
www.costofdiy.com
December 2003

# Introduction

This book is intended to help small to medium-sized plumbing businesses to carry out their estimating work more quickly and more accurately. It is hoped that the resulting saving time will enable contractors to run their businesses more efficiently.

The contents of this book cover unit rates, project costs and general advice on business matters. The unit rates section presents analytical rates for all types of work encountered in domestic and light industrial plumbing contracts. Some rates for alterations and repairs are also included.

The project costs section contains the total costs for carrying out the following types of work and each of these categories is sub-divided into different sized projects and types of materials:

- rainwater goods
- bathrooms
- external waste systems
- central heating systems
- hot and cold water systems.

The business section covers advice on starting and running a business, together with information on taxation and VAT matters.

**Rainwater goods**

This section lists the quantities and costs of PVC-U, cast iron and aluminium rainwater gutters and pipes for 18 different types and sizes of houses. These range from a one-storey terraced house, overall size 7 × 6m with gable ends, to a three-storey house, overall size 10 × 9m with hipped ends. Each house within the range is costed for PVC-U, cast iron and aluminium pipes and gutters to produce a total coverage of rainwater goods for 54 types of houses.

For example, on pages 93 and 94, the total cost of taking down the existing rainwater pipes and gutters and replacing them in PVC-U for a semi-detached two-storey house, overall size 8 × 7m with hipped ends is £528.49. This is broken down into £255.78 for labour, £203.78 for materials and £68.93 for overheads and profit. Due to rounding off, the vertical and horizontal totals may not always coincide exactly.

In the above example, the time allowed to do the work is shown as 17.64 man hours, i.e. 17 hours 38 minutes, but the actual period over which the work is carried out may be longer due to other factors. The totals of the projects are taken to a summary on pages 121 to 123.

## Bathrooms

This section covers bathroom layouts with varying combinations of the following sanitary fittings:

- bath
- bath with hand shower set
- WC
- lavatory basin
- shower cubicle
- bidet.

A total of 16 bathrooms are included and the totals are summarised on pages 161 to 162.

## External waste systems

Soil pipes and waste pipes from wash basins and baths are included covering four types of pipe for one-, two- and three-storey houses producing a combination of twelve different cases. These are summarised on page 173.

## Central heating systems

Gas-fired central heating systems for three sizes of one-, two- and three-bedroom houses are included and summarised costs appear on page 193.

## Hot and cold water systems

The project costs for the installation of hot and cold water pipes including storage tanks and cylinders, are covered for one-, two- and three-storey houses and the labour hours and costs are included on page 200.

## Materials

One of the problems facing small plumbing and heating contractors is their inability to obtain large discounts on materials because the nature of the work that they undertake precludes them from buying in large quantities. The discounts allowed in this book are generally 10-15%. The allowance for waste is 5% except for large items, such as boilers and radiators, where 1% has been included.

**Labour**

The labour rate has been set at £14.50 per hour for all work in this book and is based on current rates paid to Advanced Plumbers.

**Overheads and profit**

This has been set at 15% for all grades of work and is deemed to cover head office and site overheads, including:

- heating
- lighting
- rent
- rates
- telephones
- secretarial services
- insurances
- finance charges
- transport
- small tools
- ladders
- scaffolding etc.

# Part One

## UNIT RATES

H71 Lead sheet coverings

N13 Sanitary appliances/fittings

P31 Holes, chases, covers and supports for services

R10 Rainwater pipework/gutters

R11 Drainage above ground

R12 Drainage below ground

S10 Hot and cold water

T10 Gas/oil-fired boilers

T30 Radiators

Alterations and repairs

## H71 LEAD SHEET COVERINGS

| | Unit | Hours | Hours £ | Mat'ls £ | O & P £ | Total £ |
|---|---|---|---|---|---|---|
| Flat roofing, less than 10 degrees to the horizontal | | | | | | |
| code 4 | m2 | 2.90 | 42.05 | 17.46 | 8.93 | 68.44 |
| code 5 | m2 | 3.10 | 44.95 | 23.27 | 10.23 | 78.45 |
| code 6 | m2 | 3.30 | 47.85 | 25.70 | 11.03 | 84.58 |
| code 7 | m2 | 3.50 | 50.75 | 30.55 | 12.20 | 93.50 |
| code 8 | m2 | 3.80 | 55.10 | 34.43 | 13.43 | 102.96 |
| Dormers, less than 10 degrees to the horizontal | | | | | | |
| code 4 | m2 | 3.65 | 52.93 | 17.46 | 10.56 | 80.94 |
| code 5 | m2 | 3.90 | 56.55 | 23.27 | 11.97 | 91.79 |
| code 6 | m2 | 4.15 | 60.18 | 25.70 | 12.88 | 98.76 |
| code 7 | m2 | 4.35 | 63.08 | 30.55 | 14.04 | 107.67 |
| code 8 | m2 | 4.65 | 67.43 | 34.43 | 15.28 | 117.13 |
| Sloping roofing, 10 to 50 degrees to the horizontal | | | | | | |
| code 4 | m2 | 3.10 | 44.95 | 17.46 | 9.36 | 71.77 |
| code 5 | m2 | 3.30 | 47.85 | 23.27 | 10.67 | 81.79 |
| code 6 | m2 | 3.50 | 50.75 | 25.70 | 11.47 | 87.92 |
| code 7 | m2 | 3.70 | 53.65 | 30.55 | 12.63 | 96.83 |
| code 8 | m2 | 3.90 | 56.55 | 34.43 | 13.65 | 104.63 |
| Dormers, 10 to 50 degrees to the horizontal | | | | | | |
| code 4 | m2 | 3.65 | 52.93 | 17.46 | 10.56 | 80.94 |
| code 5 | m2 | 3.90 | 56.55 | 23.27 | 11.97 | 91.79 |
| code 6 | m2 | 4.15 | 60.18 | 25.70 | 12.88 | 98.76 |
| code 7 | m2 | 4.35 | 63.08 | 30.55 | 14.04 | 107.67 |
| code 8 | m2 | 4.65 | 67.43 | 34.43 | 15.28 | 117.13 |
| Vertical or sloping roofing, 10 to 50 degrees to the horizontal | | | | | | |
| code 4 | m2 | 3.30 | 47.85 | 17.46 | 9.80 | 75.11 |
| code 5 | m2 | 3.50 | 50.75 | 23.27 | 11.10 | 85.12 |

| | Unit | Hours | Hours £ | Mat'ls £ | O & P £ | Total £ |
|---|---|---|---|---|---|---|
| **Vertical or sloping roofing, 10 to 50 degrees to the horizontal (cont'd)** | | | | | | |
| code 6 | m2 | 3.70 | 53.65 | 25.70 | 11.90 | 91.25 |
| code 7 | m2 | 3.90 | 56.55 | 30.55 | 13.07 | 100.17 |
| code 8 | m2 | 4.10 | 59.45 | 34.43 | 14.08 | 107.96 |
| Dormers, over 50 degrees to the horizontal | | | | | | |
| code 4 | m2 | 3.65 | 52.93 | 17.46 | 10.56 | 80.94 |
| code 5 | m2 | 3.90 | 56.55 | 23.27 | 11.97 | 91.79 |
| code 6 | m2 | 4.15 | 60.18 | 25.70 | 12.88 | 98.76 |
| code 7 | m2 | 4.35 | 63.08 | 30.55 | 14.04 | 107.67 |
| code 8 | m2 | 4.65 | 67.43 | 34.43 | 15.28 | 117.13 |
| Flashings, horizontal, girth 150mm | | | | | | |
| code 4 | m | 0.30 | 4.35 | 2.62 | 1.05 | 8.02 |
| code 5 | m | 0.35 | 5.08 | 3.50 | 1.29 | 9.86 |
| Flashings, horizontal, girth 240mm | | | | | | |
| code 4 | m | 0.40 | 5.80 | 4.19 | 1.50 | 11.49 |
| code 5 | m | 0.45 | 6.53 | 5.58 | 1.82 | 13.92 |
| Flashings, horizontal, girth 300mm | | | | | | |
| code 4 | m | 0.50 | 7.25 | 5.24 | 1.87 | 14.36 |
| code 5 | m | 0.60 | 8.70 | 7.00 | 2.36 | 18.06 |
| Flashings, sloping, girth 150mm | | | | | | |
| code 4 | m | 0.40 | 5.80 | 2.62 | 1.26 | 9.68 |
| code 5 | m | 0.45 | 6.53 | 3.50 | 1.50 | 11.53 |
| Flashings, sloping, girth 240mm | | | | | | |
| code 4 | m | 0.50 | 7.25 | 4.19 | 1.72 | 13.16 |
| code 5 | m | 0.55 | 7.98 | 5.58 | 2.03 | 15.59 |
| Flashings, sloping, girth 300mm | | | | | | |
| code 4 | m | 0.60 | 8.70 | 5.24 | 2.09 | 16.03 |
| code 5 | m | 0.65 | 9.43 | 7.00 | 2.46 | 18.89 |

| | Unit | Hours | Hours £ | Mat'ls £ | O & P £ | Total £ |
|---|---|---|---|---|---|---|
| Flashings, stepped, girth 150mm | | | | | | |
| code 4 | m | 0.50 | 7.25 | 2.62 | 1.48 | 11.35 |
| code 5 | m | 0.55 | 7.98 | 3.50 | 1.72 | 13.20 |
| Flashings, stepped, girth 240mm | | | | | | |
| code 4 | m | 0.60 | 8.70 | 4.19 | 1.93 | 14.82 |
| code 5 | m | 0.65 | 9.43 | 5.58 | 2.25 | 17.26 |
| Flashings, stepped, girth 300mm | | | | | | |
| code 4 | m | 0.70 | 10.15 | 5.24 | 2.31 | 17.70 |
| code 5 | m | 0.75 | 10.88 | 7.00 | 2.68 | 20.56 |
| Aprons, horizontal, girth 240mm | | | | | | |
| code 4 | m | 0.50 | 7.25 | 4.19 | 1.72 | 13.16 |
| code 5 | m | 0.55 | 7.98 | 5.58 | 2.03 | 15.59 |
| Aprons, horizontal, girth 300mm | | | | | | |
| code 4 | m | 0.60 | 8.70 | 5.24 | 2.09 | 16.03 |
| code 5 | m | 0.65 | 9.43 | 7.00 | 2.46 | 18.89 |
| Aprons, horizontal, girth 450mm | | | | | | |
| code 4 | m | 0.70 | 10.15 | 7.86 | 2.70 | 20.71 |
| code 5 | m | 0.75 | 10.88 | 10.47 | 3.20 | 24.55 |
| Aprons, sloping, girth 240mm | | | | | | |
| code 4 | m | 0.60 | 8.70 | 4.19 | 1.93 | 14.82 |
| code 5 | m | 0.65 | 9.43 | 5.58 | 2.25 | 17.26 |
| Aprons, sloping, girth 300mm | | | | | | |
| code 4 | m | 0.70 | 10.15 | 5.24 | 2.31 | 17.70 |
| code 5 | m | 0.75 | 10.88 | 7.00 | 2.68 | 20.56 |
| Aprons, sloping, girth 450mm | | | | | | |
| code 4 | m | 0.80 | 11.60 | 7.86 | 2.92 | 22.38 |
| code 5 | m | 0.85 | 12.33 | 10.47 | 3.42 | 26.21 |
| Sills, horizontal, girth 240mm | | | | | | |
| code 4 | m | 0.50 | 7.25 | 4.19 | 1.72 | 13.16 |
| code 5 | m | 0.55 | 7.98 | 5.58 | 2.03 | 15.59 |

| | Unit | Hours | Hours £ | Mat'ls £ | O & P £ | Total £ |
|---|---|---|---|---|---|---|
| Sills, horizontal, girth 300mm | | | | | | |
| code 4 | m | 0.60 | 8.70 | 5.24 | 2.09 | 16.03 |
| code 5 | m | 0.65 | 9.43 | 7.00 | 2.46 | 18.89 |
| Sills, horizontal, girth 450mm | | | | | | |
| code 4 | m | 0.70 | 10.15 | 7.86 | 2.70 | 20.71 |
| code 5 | m | 0.75 | 10.88 | 10.47 | 3.20 | 24.55 |
| Cappings, horizontal, girth 240mm | | | | | | |
| code 4 | m | 0.50 | 7.25 | 4.19 | 1.72 | 13.16 |
| code 5 | m | 0.55 | 7.98 | 5.58 | 2.03 | 15.59 |
| Cappings, horizontal, girth 300mm | | | | | | |
| code 4 | m | 0.70 | 10.15 | 5.24 | 2.31 | 17.70 |
| code 5 | m | 0.80 | 11.60 | 7.00 | 2.79 | 21.39 |
| Cappings, horizontal, girth 450mm | | | | | | |
| code 4 | m | 1.00 | 14.50 | 7.86 | 3.35 | 25.71 |
| code 5 | m | 1.10 | 15.95 | 10.47 | 3.96 | 30.38 |
| Hips, sloping, girth 240mm | | | | | | |
| code 4 | m | 0.55 | 7.98 | 4.19 | 1.82 | 13.99 |
| code 5 | m | 0.65 | 9.43 | 5.58 | 2.25 | 17.26 |
| Hips, sloping, girth 300mm | | | | | | |
| code 4 | m | 0.80 | 11.60 | 5.24 | 2.53 | 19.37 |
| code 5 | m | 0.90 | 13.05 | 7.00 | 3.01 | 23.06 |
| Hips, sloping, girth 450mm | | | | | | |
| code 4 | m | 1.10 | 15.95 | 7.86 | 3.57 | 27.38 |
| code 5 | m | 1.20 | 17.40 | 10.47 | 4.18 | 32.05 |
| Kerbs, horizontal, girth 240mm | | | | | | |
| code 4 | m | 0.60 | 8.70 | 4.19 | 1.93 | 14.82 |
| code 5 | m | 0.70 | 10.15 | 5.58 | 2.36 | 18.09 |
| Kerbs, horizontal, girth 300mm | | | | | | |
| code 4 | m | 0.70 | 10.15 | 5.24 | 2.31 | 17.70 |
| code 5 | m | 0.80 | 11.60 | 7.00 | 2.79 | 21.39 |

| | Unit | Hours | Hours £ | Mat'ls £ | O & P £ | Total £ |
|---|---|---|---|---|---|---|
| Kerbs, horizontal, girth 450mm | | | | | | |
| code 4 | m | 1.00 | 14.50 | 7.86 | 3.35 | 25.71 |
| code 5 | m | 1.10 | 15.95 | 10.47 | 3.96 | 30.38 |
| Ridges, horizontal, girth 240mm | | | | | | |
| code 4 | m | 0.60 | 8.70 | 4.19 | 1.93 | 14.82 |
| code 5 | m | 0.70 | 10.15 | 5.58 | 2.36 | 18.09 |
| Ridges, horizontal, girth 300mm | | | | | | |
| code 4 | m | 0.70 | 10.15 | 5.24 | 2.31 | 17.70 |
| code 5 | m | 0.80 | 11.60 | 7.00 | 2.79 | 21.39 |
| Ridges, horizontal, girth 450mm | | | | | | |
| code 4 | m | 1.00 | 14.50 | 7.86 | 3.35 | 25.71 |
| code 5 | m | 1.10 | 15.95 | 10.47 | 3.96 | 30.38 |
| Valleys, horizontal, girth 450mm | | | | | | |
| code 4 | m | 1.00 | 14.50 | 7.86 | 3.35 | 25.71 |
| code 5 | m | 1.10 | 15.95 | 10.47 | 3.96 | 30.38 |
| Valleys, horizontal, girth 600mm | | | | | | |
| code 4 | m | 1.20 | 17.40 | 10.48 | 4.18 | 32.06 |
| code 5 | m | 1.30 | 18.85 | 13.96 | 4.92 | 37.73 |
| Gutters, sloping, girth 450mm | | | | | | |
| code 4 | m | 1.10 | 15.95 | 7.86 | 3.57 | 27.38 |
| code 5 | m | 1.30 | 18.85 | 10.47 | 4.40 | 33.72 |
| code 6 | m | 1.50 | 21.75 | 11.56 | 5.00 | 38.31 |
| code 7 | m | 1.60 | 23.20 | 13.74 | 5.54 | 42.48 |
| Gutters, sloping, girth 600mm | | | | | | |
| code 4 | m | 1.30 | 18.85 | 10.48 | 4.40 | 33.73 |
| code 5 | m | 1.50 | 21.75 | 13.96 | 5.36 | 41.07 |
| code 6 | m | 1.70 | 24.65 | 15.42 | 6.01 | 46.08 |
| code 7 | m | 1.80 | 26.10 | 18.33 | 6.66 | 51.09 |
| Edges, welted | | | | | | |
| code 4 | m | 0.30 | 4.35 | 0.00 | 0.65 | 5.00 |
| code 5 | m | 0.30 | 4.35 | 0.00 | 0.65 | 5.00 |
| code 6 | m | 0.50 | 7.25 | 0.00 | 1.09 | 8.34 |

| | Unit | Hours | Hours £ | Mat'ls £ | O & P £ | Total £ |
|---|---|---|---|---|---|---|
| Edges, beaded | | | | | | |
| code 4 | m | 0.30 | 4.35 | 0.00 | 0.65 | 5.00 |
| code 5 | m | 0.30 | 4.35 | 0.00 | 0.65 | 5.00 |
| code 6 | m | 0.50 | 7.25 | 0.00 | 1.09 | 8.34 |
| Dressing over slating and tiling | | | | | | |
| code 4 | m | 0.20 | 2.90 | 0.00 | 0.44 | 3.34 |
| code 5 | m | 0.25 | 3.63 | 0.00 | 0.54 | 4.17 |
| code 6 | m | 0.30 | 4.35 | 0.00 | 0.65 | 5.00 |
| code 7 | m | 0.35 | 5.08 | 0.00 | 0.76 | 5.84 |
| Soakers, size 150 × 150mm | | | | | | |
| code 3 | nr | 0.20 | 2.90 | 1.10 | 0.60 | 4.60 |
| code 4 | nr | 0.25 | 3.63 | 1.24 | 0.73 | 5.59 |
| Soakers, size 200 × 300mm | | | | | | |
| code 3 | nr | 0.25 | 3.63 | 1.85 | 0.82 | 6.30 |
| code 4 | nr | 0.30 | 4.35 | 2.14 | 0.97 | 7.46 |
| Soakers, size 300 × 150mm | | | | | | |
| code 3 | nr | 0.30 | 4.35 | 2.22 | 0.99 | 7.56 |
| code 4 | nr | 0.35 | 5.08 | 2.56 | 1.15 | 8.78 |
| Slates, size 400 × 400mm with 200mm high collar 100mm diameter | | | | | | |
| code 4 | nr | 1.40 | 20.30 | 8.88 | 4.38 | 33.56 |
| code 5 | nr | 1.50 | 21.75 | 9.57 | 4.70 | 36.02 |
| Slates, size 400 × 400mm with 200mm high collar 150mm diameter | | | | | | |
| code 4 | nr | 1.60 | 23.20 | 9.32 | 4.88 | 37.40 |
| code 5 | nr | 1.70 | 24.65 | 9.64 | 5.14 | 39.43 |
| Dots, cast lead | | | | | | |
| code 3 | nr | 0.70 | 10.15 | 1.48 | 1.74 | 13.37 |
| code 4 | nr | 0.75 | 10.88 | 1.60 | 1.87 | 14.35 |
| code 5 | nr | 0.80 | 11.60 | 1.72 | 2.00 | 15.32 |

| | Unit | Hours | Hours £ | Mat'ls £ | O & P £ | Total £ |
|---|---|---|---|---|---|---|
| **N13 SANITARY APPLIANCES/ FITTINGS** | | | | | | |
| **Baths** | | | | | | |
| Acrylic reinforced bath size 1500 × 700mm complete with 2nr chromium plated grips, 40mm waste fitting, overflow with chain and plastic plug | | | | | | |
| white | nr | 2.75 | 39.88 | 124.56 | 24.67 | 189.10 |
| Acrylic reinforced bath size 1700 × 700mm complete with 2nr chromium plated grips, 40mm waste fitting, overflow with chain and plastic plug | | | | | | |
| white | nr | 2.90 | 42.05 | 141.23 | 27.49 | 210.77 |
| coloured | nr | 2.90 | 42.05 | 153.81 | 29.38 | 225.24 |
| Acrylic reinforced bath size 1700 × 800mm complete with 2nr chromium plated grips, 40mm waste fitting, overflow with chain and plastic plug | | | | | | |
| white | nr | 3.00 | 43.50 | 171.24 | 32.21 | 246.95 |
| coloured | nr | 3.00 | 43.50 | 187.64 | 34.67 | 265.81 |
| Acrylic reinforced bath size 1700 × 800mm complete with 2nr chromium plated grips, 40mm waste fitting, overflow with chain and plastic plug | | | | | | |
| white | nr | 3.10 | 44.95 | 221.47 | 39.96 | 306.38 |
| coloured | nr | 3.10 | 44.95 | 231.37 | 41.45 | 317.77 |

| | Unit | Hours | Hours £ | Mat'ls £ | O & P £ | Total £ |
|---|---|---|---|---|---|---|
| Porcelain enamel standard gauge bath size 1700 × 700mm complete with 2nr chromium plated grips, 40mm waste fitting, overflow with chain and plastic plug | | | | | | |
| white | nr | 3.00 | 43.50 | 147.94 | 28.72 | 220.16 |
| coloured | nr | 3.00 | 43.50 | 161.28 | 30.72 | 235.50 |
| Porcelain enamel heavy gauge bath size 1700 × 700mm complete with 2nr chromium plated grips, 40mm waste fitting, overflow with chain and plastic plug | | | | | | |
| white | nr | 3.10 | 44.95 | 154.67 | 29.94 | 229.56 |
| coloured | nr | 3.10 | 44.95 | 194.69 | 35.95 | 275.59 |
| Porcelain enamel heavy gauge (shallow) bath size 1700 × 700mm complete with 2nr chromium plated grips, 40mm waste fitting, overflow with chain and plastic plug | | | | | | |
| white | nr | 3.10 | 44.95 | 228.14 | 40.96 | 314.05 |
| coloured | nr | 3.10 | 44.95 | 265.54 | 46.57 | 357.06 |
| **Bath accessories** | | | | | | |
| Bath panels, enamelled hardboard, fixing with chromium-plated dome headed screws, cutting to length | | | | | | |
| end panel | nr | 0.30 | 4.35 | 9.87 | 2.13 | 16.35 |
| side panel | nr | 0.50 | 7.25 | 11.89 | 2.87 | 22.01 |
| Bath panels, moulded acrylic, fixing in position for trimming as required | | | | | | |
| end panel | nr | 0.30 | 4.35 | 12.55 | 2.54 | 19.44 |
| side panel | nr | 0.50 | 7.25 | 14.99 | 3.34 | 25.58 |

| | Unit | Hours | Hours £ | Mat'ls £ | O & P £ | Total £ |
|---|---|---|---|---|---|---|
| Angle strip polished aluminium, fixing with chromium-plated dome headed screws, cutting to length | | | | | | |
| 25 × 25 × 560mm long | nr | 0.35 | 5.08 | 4.55 | 1.44 | 11.07 |
| **Basins and pedestals** | | | | | | |
| Wash basin, vitreous china size 510 × 410mm complete with pedestal, chromium-plated waste, overflow with chain and plastic plug | | | | | | |
| white | nr | 2.20 | 31.90 | 82.46 | 17.15 | 131.51 |
| coloured | nr | 2.20 | 31.90 | 89.75 | 18.25 | 139.90 |
| Wash basin, vitreous china size 560 × 455mm complete with pedestal, chromium-plated waste, overflow with chain and plastic plug | | | | | | |
| white | nr | 2.25 | 32.63 | 75.65 | 16.24 | 124.52 |
| coloured | nr | 2.25 | 32.63 | 91.55 | 18.63 | 142.80 |
| Wash basin, vitreous china size 610 × 470mm complete with pedestal, chromium-plated waste, overflow with chain and plastic plug | | | | | | |
| white | nr | 2.30 | 33.35 | 85.61 | 17.84 | 136.80 |
| coloured | nr | 2.30 | 33.35 | 92.55 | 18.89 | 144.79 |
| Wash basin, vitreous china size 510 × 410mm complete with brackets, chromium-plated waste, overflow with chain and plastic plug | | | | | | |
| white | nr | 2.00 | 29.00 | 63.51 | 13.88 | 106.39 |
| coloured | nr | 2.00 | 29.00 | 72.92 | 15.29 | 117.21 |

| | Unit | Hours | Hours £ | Mat'ls £ | O & P £ | Total £ |
|---|---|---|---|---|---|---|
| Wash basin, vitreous china size 560 × 455mm complete with brackets, chromium-plated waste, overflow with chain and plastic plug | | | | | | |
| white | nr | 2.25 | 32.63 | 65.33 | 14.69 | 112.65 |
| coloured | nr | 2.25 | 32.63 | 75.61 | 16.24 | 124.47 |
| Wash basin, vitreous china size 610 × 470mm complete with brackets, chromium-plated waste, overflow with chain and plastic plug | | | | | | |
| white | nr | 2.30 | 33.35 | 68.94 | 15.34 | 117.63 |
| coloured | nr | 2.30 | 33.35 | 75.63 | 16.35 | 125.33 |
| **Sinks and tops** | | | | | | |
| Stainless steel single sit-on type sink, complete with inset chromium-plated waste, overflow with chain and plastic plug, size | | | | | | |
| 1000 × 500mm | nr | 1.30 | 18.85 | 86.24 | 15.76 | 120.85 |
| 1000 × 600mm | nr | 1.40 | 20.30 | 92.65 | 16.94 | 129.89 |
| 1200 × 600mm | nr | 1.50 | 21.75 | 104.22 | 18.90 | 144.87 |
| Stainless steel single roll-edge type sink, complete with inset chromium-plated waste, overflow with chain and plastic plug, size | | | | | | |
| 1000 × 500mm | nr | 1.30 | 18.85 | 89.05 | 16.19 | 124.09 |
| 1000 × 600mm | nr | 1.40 | 20.30 | 96.61 | 17.54 | 134.45 |
| 1200 × 600mm | nr | 1.50 | 21.75 | 108.04 | 19.47 | 149.26 |
| Stainless steel double sit-on type sink, complete with inset chromium-plated waste, overflow with chain and plastic plug, size | | | | | | |
| 1500 × 500mm | nr | 1.80 | 26.10 | 107.54 | 20.05 | 153.69 |
| 1500 × 600mm | nr | 1.90 | 27.55 | 123.64 | 22.68 | 173.87 |

| | Unit | Hours | Hours £ | Mat'ls £ | O & P £ | Total £ |
|---|---|---|---|---|---|---|
| Stainless steel double roll-edge type sink, complete with inset chromium-plated waste, overflow with chain and plastic plug, size | | | | | | |
| 1000 × 500mm | nr | 1.80 | 26.10 | 115.71 | 21.27 | 163.08 |
| 1000 × 600mm | nr | 1.90 | 27.55 | 128.61 | 23.42 | 179.58 |
| Belfast pattern white fireclay sink, complete with chromium-plated waste,chain and plastic plug, wall-mounted on brackets, size | | | | | | |
| 450 × 380 × 205mm | nr | 2.35 | 34.08 | 115.57 | 22.45 | 172.09 |
| 610 × 455 × 255mm | nr | 2.40 | 34.80 | 161.23 | 29.40 | 225.43 |
| 760 × 455 × 255mm | nr | 2.45 | 26.80 | 215.66 | 36.37 | 278.83 |
| **WC suites** | | | | | | |
| Vitreous china low-level WC suite comprising pan, plastic seat and cover, 9 litre cistern, low-pressure ball valve, connecting pipework screwed to floor | nr | 2.35 | 34.08 | 204.56 | 35.80 | 274.43 |
| Vitreous china low-level close coupled WC suite comprising pan, plastic seat and cover, 9 litre cistern, low-pressure ball valve, connecting pipework screwed to floor | nr | 2.45 | 35.53 | 274.97 | 46.57 | 357.07 |
| **Bidets** | | | | | | |
| Free-standing plain rim, vitreous china bidet excluding fittings | | | | | | |
| white | nr | 2.50 | 36.25 | 258.79 | 44.26 | 339.30 |
| coloured | nr | 2.50 | 36.25 | 278.40 | 47.20 | 361.85 |

| | Unit | Hours | Hours £ | Mat'ls £ | O & P £ | Total £ |
|---|---|---|---|---|---|---|
| **Urinals** | | | | | | |
| Bowl urinals, white vitreous china, wall-mounted on hangers chromium-plated dome outlet | | | | | | |
| 430mm wide × 305mm high | nr | 1.00 | 14.50 | 110.14 | 18.70 | 143.34 |
| 500mm wide × 350mm high | nr | 1.10 | 15.95 | 152.67 | 25.29 | 193.91 |
| Stainless steel flush pipes and spreaders for bowl urinals, face fixed to wall | | | | | | |
| one bowl set | nr | 1.20 | 17.40 | 58.97 | 11.46 | 87.83 |
| two bowl set | nr | 1.75 | 25.38 | 102.46 | 19.18 | 147.01 |
| three bowl set | nr | 2.10 | 30.45 | 148.92 | 26.91 | 206.28 |
| Automatic flushing cistern and fittings, white vitreous china, wall-mounted on brackets | | | | | | |
| 4.5 litre | nr | 1.00 | 14.50 | 124.76 | 20.89 | 160.15 |
| 9 litre | nr | 1.05 | 15.23 | 140.27 | 23.32 | 178.82 |
| 13.5 litre | nr | 1.15 | 16.68 | 169.67 | 27.95 | 214.30 |
| Modular slab urinal, white fireclay china, comprising back slabs without divisions, waterway channel, automatic flushing cistern, stainless steel flush pipes | | | | | | |
| two persons | nr | 5.20 | 75.40 | 602.27 | 101.65 | 779.32 |
| three persons | nr | 6.00 | 87.00 | 987.68 | 161.20 | 1235.88 |
| four persons | nr | 7.00 | 101.50 | 1250.67 | 202.83 | 1555.00 |
| **Taps** | | | | | | |
| Chromium-plated pillar taps | | | | | | |
| 13mm | pr | 0.40 | 5.80 | 23.88 | 4.45 | 34.13 |
| 19mm | pr | 0.45 | 6.53 | 31.92 | 5.77 | 44.21 |

| | Unit | Hours | Hours £ | Mat'ls £ | O & P £ | Total £ |
|---|---|---|---|---|---|---|
| Chromium-plated mixer taps, 19mm | | | | | | |
| cross-top handles | pr | 0.60 | 8.70 | 58.93 | 10.14 | 77.77 |
| lever handles | pr | 0.60 | 8.70 | 79.38 | 13.21 | 101.29 |
| Chromium-plated bidet monoblock and spray | nr | 0.70 | 10.15 | 170.26 | 27.06 | 207.47 |
| **Valves** | | | | | | |
| Chromium-plated thermostatic exposed shower valve with flexible hose, slide rail and spray set | nr | 1.40 | 20.30 | 201.56 | 33.28 | 255.14 |
| Instant electric 9.5kw shower with flexible hose, slide rail and spray set | nr | 1.50 | 21.75 | 167.28 | 28.35 | 217.38 |
| **Showers** | | | | | | |
| Shower cubicle size 788 × 842 × 2115mm in anodised aluminium frame and safety glass | nr | 2.00 | 29.00 | 689.48 | 107.77 | 826.25 |
| White glazed fireclay shower tray size 900 × 900 × 180mm with chromium-plated waste fitting | nr | 1.80 | 26.10 | 178.71 | 30.72 | 235.53 |
| White acrylic fireclay shower tray size 750 × 750 × 180mm with chromium-plated waste fitting | nr | 1.30 | 18.85 | 94.35 | 16.98 | 130.18 |

| | Unit | Hours | Hours £ | Mat'ls £ | O & P £ | Total £ |
|---|---|---|---|---|---|---|
| **P31 HOLES, CHASES, COVERS AND SUPPORTS FOR SERVICES** | | | | | | |
| Cutting holes for pipes up to 25mm diameter in walls 102.5mm thick | | | | | | |
| commons | nr | 0.40 | 5.80 | 0.00 | 0.87 | 6.67 |
| facings | nr | 0.70 | 10.15 | 0.00 | 1.52 | 11.67 |
| engineering class A | nr | 0.90 | 13.05 | 0.00 | 1.96 | 15.01 |
| engineering class B | nr | 0.75 | 10.88 | 0.00 | 1.63 | 12.51 |
| Cutting holes for pipes up to 25mm diameter in walls 100mm thick | | | | | | |
| concrete blocks | nr | 0.30 | 4.35 | 0.00 | 0.65 | 5.00 |
| Thermalite blocks | nr | 0.20 | 2.90 | 0.00 | 0.44 | 3.34 |
| aerated concrete blocks | nr | 0.25 | 3.63 | 0.00 | 0.54 | 4.17 |
| Cutting holes for pipes up to 25mm diameter in walls 225mm thick | | | | | | |
| commons | nr | 0.80 | 11.60 | 0.00 | 1.74 | 13.34 |
| facings | nr | 0.85 | 12.33 | 0.00 | 1.85 | 14.17 |
| engineering class A | nr | 1.30 | 18.85 | 0.00 | 2.83 | 21.68 |
| engineering class B | nr | 1.50 | 21.75 | 0.00 | 3.26 | 25.01 |
| Cutting holes for pipes up to 25mm diameter in walls 100mm thick | | | | | | |
| concrete blocks | nr | 0.45 | 6.53 | 0.00 | 0.98 | 7.50 |
| Thermalite blocks | nr | 0.30 | 4.35 | 0.00 | 0.65 | 5.00 |
| aerated concrete blocks | nr | 0.35 | 5.08 | 0.00 | 0.76 | 5.84 |

| | Unit | Hours | Hours £ | Mat'ls £ | O & P £ | Total £ |
|---|---|---|---|---|---|---|
| Cutting holes for pipes up to 25 to 50mm diameter in walls 102.5mm thick | | | | | | |
| commons | nr | 0.50 | 7.25 | 0.00 | 1.09 | 8.34 |
| facings | nr | 0.80 | 11.60 | 0.00 | 1.74 | 13.34 |
| engineering class A | nr | 1.00 | 14.50 | 0.00 | 2.18 | 16.68 |
| engineering class B | nr | 0.85 | 12.33 | 0.00 | 1.85 | 14.17 |
| Cutting holes for pipes up to 25 to 50mm diameter in walls 100mm thick | | | | | | |
| concrete blocks | nr | 0.40 | 5.80 | 0.00 | 0.87 | 6.67 |
| Thermalite blocks | nr | 0.30 | 4.35 | 0.00 | 0.65 | 5.00 |
| aerated concrete blocks | nr | 0.35 | 5.08 | 0.00 | 0.76 | 5.84 |
| Cutting holes for pipes up to 25 to 50mm diameter in walls 225mm thick | | | | | | |
| commons | nr | 1.05 | 15.23 | 0.00 | 2.28 | 17.51 |
| facings | nr | 1.10 | 15.95 | 0.00 | 2.39 | 18.34 |
| engineering class A | nr | 1.60 | 23.20 | 0.00 | 3.48 | 26.68 |
| engineering class B | nr | 1.80 | 26.10 | 0.00 | 3.92 | 30.02 |
| Cutting holes for pipes up to 25 to 50mm diameter in walls 100mm thick | | | | | | |
| concrete blocks | nr | 0.60 | 8.70 | 0.00 | 1.31 | 10.01 |
| Thermalite blocks | nr | 0.40 | 5.80 | 0.00 | 0.87 | 6.67 |
| aerated concrete blocks | nr | 0.45 | 6.53 | 0.00 | 0.98 | 7.50 |
| Cutting and pinning ends of pipe support brackets and make good to | | | | | | |
| concrete | nr | 0.45 | 6.53 | 0.00 | 0.98 | 7.50 |
| brickwork | nr | 0.35 | 5.08 | 0.00 | 0.76 | 5.84 |
| blockwork | nr | 0.30 | 4.35 | 0.00 | 0.65 | 5.00 |
| tiled walls | nr | 0.50 | 7.25 | 0.00 | 1.09 | 8.34 |

| | Unit | Hours | Hours £ | Mat'ls £ | O & P £ | Total £ |
|---|---|---|---|---|---|---|
| Cut opening through cavity wall comprising facing bricks and concrete blocks | | | | | | |
| balanced flue outlet | nr | 0.90 | 13.05 | 0.00 | 1.96 | 15.01 |
| 150mm diameter flue pipe | nr | 0.75 | 10.88 | 0.00 | 1.63 | 12.51 |
| 19mm overflow pipe | nr | 0.20 | 2.90 | 0.00 | 0.44 | 3.34 |
| Cutting chases for one pipe up to 25mm diameter in | | | | | | |
| commons | nr | 0.35 | 5.08 | 0.00 | 0.76 | 5.84 |
| facings | nr | 0.38 | 5.51 | 0.00 | 0.83 | 6.34 |
| engineering class A | nr | 0.45 | 6.53 | 0.00 | 0.98 | 7.50 |
| engineering class B | nr | 0.40 | 5.80 | 0.00 | 0.87 | 6.67 |
| concrete blocks | nr | 0.35 | 5.08 | 0.00 | 0.76 | 5.84 |
| Thermalite blocks | nr | 0.25 | 3.63 | 0.00 | 0.54 | 4.17 |
| aerated concrete blocks | nr | 0.20 | 2.90 | 0.00 | 0.44 | 3.34 |
| Cutting chases for two pipes up to 25mm diameter or one pipe 50mm diameter in | | | | | | |
| commons | m | 0.45 | 6.53 | 0.00 | 0.98 | 7.50 |
| facings | m | 0.50 | 7.25 | 0.00 | 1.09 | 8.34 |
| engineering class A | m | 0.65 | 9.43 | 0.00 | 1.41 | 10.84 |
| engineering class B | m | 0.55 | 7.98 | 0.00 | 1.20 | 9.17 |
| concrete blocks | m | 0.40 | 5.80 | 0.00 | 0.87 | 6.67 |
| Thermalite blocks | m | 0.30 | 4.35 | 0.00 | 0.65 | 5.00 |
| aerated concrete blocks | m | 0.90 | 13.05 | 0.00 | 1.96 | 15.01 |
| Make good surfaces both sides of chases | | | | | | |
| plastered surfaces | m | 0.15 | 2.18 | 0.65 | 0.42 | 3.25 |
| tiled surfaces | m | 0.25 | 3.63 | 2.15 | 0.87 | 6.64 |
| concrete floor | m | 0.35 | 5.08 | 0.65 | 0.86 | 6.58 |
| granolithic floor | m | 0.50 | 7.25 | 0.70 | 1.19 | 9.14 |
| concrete soffit | m | 0.55 | 7.98 | 0.65 | 1.29 | 9.92 |

## R10 RAINWATER PIPEWORK/GUTTERS

### PVC-U rainwater pipe plugged to brickwork with pipe brackets and fitting clips at 2m maximum centres

| | Unit | Hours | Hours £ | Mat'ls £ | O & P £ | Total £ |
|---|---|---|---|---|---|---|
| 68mm diameter pipe | m | 0.25 | 3.63 | 5.61 | 1.39 | 10.62 |
| extra over for | | | | | | |
| bend, 87.5 degrees | nr | 0.25 | 3.63 | 5.39 | 1.35 | 10.37 |
| offset | nr | 0.25 | 3.63 | 3.21 | 1.03 | 7.86 |
| branch | nr | 0.25 | 3.63 | 10.73 | 2.15 | 16.51 |
| shoe | nr | 0.25 | 3.63 | 4.65 | 1.24 | 9.52 |
| access pipe | nr | 0.25 | 3.63 | 14.77 | 2.76 | 21.15 |
| hopper head | nr | 0.25 | 3.63 | 17.26 | 3.13 | 24.02 |
| 68mm square pipe | m | 0.25 | 3.63 | 6.72 | 1.55 | 11.90 |
| extra over for | | | | | | |
| bend, 87.5 degrees | nr | 0.25 | 3.63 | 3.44 | 1.06 | 8.12 |
| offset | nr | 0.25 | 3.63 | 6.04 | 1.45 | 11.11 |
| branch | nr | 0.25 | 3.63 | 12.16 | 2.37 | 18.15 |
| shoe | nr | 0.25 | 3.63 | 3.95 | 1.14 | 8.71 |
| access pipe | nr | 0.25 | 3.63 | 18.44 | 3.31 | 25.37 |
| hopper head | nr | 0.25 | 3.63 | 17.26 | 3.13 | 24.02 |

### Aluminium rainwater pipe, straight, plain eared, spigot and socket dry joints plugged to brickwork

| | Unit | Hours | Hours £ | Mat'ls £ | O & P £ | Total £ |
|---|---|---|---|---|---|---|
| 63mm diameter pipe | m | 0.28 | 4.06 | 15.33 | 2.91 | 22.30 |
| extra over for | | | | | | |
| offset, 75mm | nr | 0.28 | 4.06 | 29.01 | 4.96 | 38.03 |
| offset, 100mm | nr | 0.28 | 4.06 | 29.01 | 4.96 | 38.03 |
| offset, 150mm | nr | 0.28 | 4.06 | 29.01 | 4.96 | 38.03 |
| offset, 225mm | nr | 0.28 | 4.06 | 34.82 | 5.83 | 44.71 |
| offset, 300mm | nr | 0.28 | 4.06 | 34.82 | 5.83 | 44.71 |

| | Unit | Hours | Hours £ | Mat'ls £ | O & P £ | Total £ |
|---|---|---|---|---|---|---|
| **63mm diameter pipe (cont'd)** | | | | | | |
| bend, 92.5 degrees | nr | 0.28 | 4.06 | 23.61 | 4.15 | 31.82 |
| branch | nr | 0.28 | 4.06 | 25.26 | 4.40 | 33.72 |
| shoe | nr | 0.28 | 4.06 | 8.80 | 1.93 | 14.79 |
| standard hopper | nr | 0.28 | 4.06 | 24.53 | 4.29 | 32.88 |
| flatback hopper | nr | 0.28 | 4.06 | 17.14 | 3.18 | 24.38 |
| 76mm diameter pipe | m | 0.30 | 4.35 | 18.89 | 3.49 | 26.73 |
| extra over for | | | | | | |
| offset, 75mm | nr | 0.30 | 4.35 | 35.69 | 6.01 | 46.05 |
| offset, 100mm | nr | 0.30 | 4.35 | 35.69 | 6.01 | 46.05 |
| offset, 150mm | nr | 0.30 | 4.35 | 35.69 | 6.01 | 46.05 |
| offset, 225mm | nr | 0.30 | 4.35 | 42.27 | 6.99 | 53.61 |
| offset, 300mm | nr | 0.30 | 4.35 | 42.82 | 7.08 | 54.25 |
| bend, 92.5 degrees | nr | 0.30 | 4.35 | 26.54 | 4.63 | 35.52 |
| branch | nr | 0.30 | 4.35 | 30.35 | 5.21 | 39.91 |
| shoe | nr | 0.30 | 4.35 | 12.16 | 2.48 | 18.99 |
| standard hopper | nr | 0.30 | 4.35 | 25.66 | 4.50 | 34.51 |
| flatback hopper | nr | 0.30 | 4.35 | 17.14 | 3.22 | 24.71 |
| 102mm diameter pipe | m | 0.32 | 4.64 | 26.38 | 4.65 | 35.67 |
| extra over for | | | | | | |
| offset, 75mm | nr | 0.32 | 4.64 | 43.39 | 7.20 | 55.23 |
| offset, 100mm | nr | 0.32 | 4.64 | 43.39 | 7.20 | 55.23 |
| offset, 150mm | nr | 0.32 | 4.64 | 43.39 | 7.20 | 55.23 |
| offset, 225mm | nr | 0.32 | 4.64 | 50.63 | 8.29 | 63.56 |
| offset, 300mm | nr | 0.32 | 4.64 | 50.63 | 8.29 | 63.56 |
| bend, 92.5 degrees | nr | 0.32 | 4.64 | 38.05 | 6.40 | 49.09 |
| branch | nr | 0.32 | 4.64 | 40.81 | 6.82 | 52.27 |
| shoe | nr | 0.32 | 4.64 | 16.05 | 3.10 | 23.79 |
| standard hopper | nr | 0.32 | 4.64 | 34.35 | 5.85 | 44.84 |
| ornamental hopper | nr | 0.32 | 4.64 | 120.00 | 18.70 | 143.34 |

| | Unit | Hours | Hours £ | Mat'ls £ | O & P £ | Total £ |
|---|---|---|---|---|---|---|
| **Cast iron rainwater pipe, straight, ears cast on, spigot and socket dry joints plugged to brickwork** | | | | | | |
| 65mm diameter pipe | m | 0.25 | 3.63 | 20.47 | 3.61 | 27.71 |
| extra over for | | | | | | |
| offset, 75mm | nr | 0.25 | 3.63 | 15.28 | 2.84 | 21.74 |
| offset, 115mm | nr | 0.25 | 3.63 | 15.28 | 2.84 | 21.74 |
| offset, 150mm | nr | 0.25 | 3.63 | 15.28 | 2.84 | 21.74 |
| offset, 225mm | nr | 0.25 | 3.63 | 17.79 | 3.21 | 24.63 |
| offset, 305mm | nr | 0.25 | 3.63 | 20.83 | 3.67 | 28.12 |
| bend | nr | 0.25 | 3.63 | 9.98 | 2.04 | 15.65 |
| branch | nr | 0.25 | 3.63 | 19.62 | 3.49 | 26.73 |
| shoe | nr | 0.25 | 3.63 | 14.14 | 2.66 | 20.43 |
| eared shoe | nr | 0.25 | 3.63 | 16.30 | 2.99 | 22.91 |
| flat hopper | nr | 0.25 | 3.63 | 12.74 | 2.45 | 18.82 |
| rectangular hopper | nr | 0.25 | 3.63 | 56.74 | 9.05 | 69.42 |
| 75mm diameter pipe | m | 0.30 | 4.35 | 20.51 | 3.73 | 28.59 |
| extra over for | | | | | | |
| offset, 75mm | nr | 0.30 | 4.35 | 15.28 | 2.94 | 22.57 |
| offset, 115mm | nr | 0.30 | 4.35 | 15.28 | 2.94 | 22.57 |
| offset, 150mm | nr | 0.30 | 4.35 | 15.28 | 2.94 | 22.57 |
| offset, 225mm | nr | 0.30 | 4.35 | 17.79 | 3.32 | 25.46 |
| offset, 305mm | nr | 0.30 | 4.35 | 21.86 | 3.93 | 30.14 |
| bend | nr | 0.30 | 4.35 | 12.12 | 2.47 | 18.94 |
| branch | nr | 0.30 | 4.35 | 21.22 | 3.84 | 29.41 |
| shoe | nr | 0.30 | 4.35 | 14.14 | 2.77 | 21.26 |
| eared shoe | nr | 0.30 | 4.35 | 16.30 | 3.10 | 23.75 |
| flat hopper | nr | 0.30 | 4.35 | 14.47 | 2.82 | 21.64 |
| rectangular hopper | nr | 0.30 | 4.35 | 56.74 | 9.16 | 70.25 |
| 100mm diameter pipe | m | 0.35 | 5.08 | 26.80 | 4.78 | 36.66 |
| extra over for | | | | | | |
| offset, 75mm | nr | 0.35 | 5.08 | 28.83 | 5.09 | 38.99 |
| offset, 115mm | nr | 0.35 | 5.08 | 29.39 | 5.17 | 39.63 |
| offset, 150mm | nr | 0.35 | 5.08 | 28.83 | 5.09 | 38.99 |
| offset, 225mm | nr | 0.35 | 5.08 | 34.91 | 6.00 | 45.98 |
| offset, 305mm | nr | 0.35 | 5.08 | 56.74 | 9.27 | 71.09 |
| bend | nr | 0.35 | 5.08 | 17.12 | 3.33 | 25.52 |

| | Unit | Hours | Hours £ | Mat'ls £ | O & P £ | Total £ |
|---|---|---|---|---|---|---|
| **100mm diameter pipe (cont'd)** | | | | | | |
| branch | nr | 0.35 | 5.08 | 25.21 | 4.54 | 34.83 |
| shoe | nr | 0.35 | 5.08 | 19.06 | 3.62 | 27.76 |
| eared shoe | nr | 0.35 | 5.08 | 21.66 | 4.01 | 30.75 |
| flat hopper | nr | 0.35 | 5.08 | 32.06 | 5.57 | 42.71 |
| rectangular hopper | nr | 0.35 | 5.08 | 56.74 | 9.27 | 71.09 |
| **PVC-U half round rainwater gutter, fixed to timber with support brackets at 1m maximum centres** | | | | | | |
| 76mm elliptical gutter | m | 0.20 | 2.90 | 6.07 | 1.35 | 10.32 |
| extra over for | | | | | | |
| running outlet | nr | 0.20 | 2.90 | 5.53 | 1.26 | 9.69 |
| angle | nr | 0.20 | 2.90 | 6.03 | 1.34 | 10.27 |
| stop end outlet | nr | 0.10 | 1.45 | 4.85 | 0.95 | 7.25 |
| stop end | nr | 0.10 | 1.45 | 2.97 | 0.66 | 5.08 |
| 112mm half-round gutter | m | 0.26 | 3.77 | 4.48 | 1.24 | 9.49 |
| extra over for | | | | | | |
| running outlet | nr | 0.26 | 3.77 | 3.90 | 1.15 | 8.82 |
| angle | nr | 0.26 | 3.77 | 4.45 | 1.23 | 9.45 |
| stop end outlet | nr | 0.13 | 1.89 | 3.90 | 0.87 | 6.65 |
| stop end | nr | 0.13 | 1.89 | 1.95 | 0.58 | 4.41 |
| 150mm half-round gutter | m | 0.30 | 4.35 | 6.61 | 1.64 | 12.60 |
| extra over for | | | | | | |
| running outlet | nr | 0.30 | 4.35 | 9.00 | 2.00 | 15.35 |
| angle | nr | 0.30 | 4.35 | 9.27 | 2.04 | 15.66 |
| stop end outlet | nr | 0.15 | 2.18 | 5.66 | 1.18 | 9.01 |
| stop end | nr | 0.15 | 2.18 | 1.82 | 0.60 | 4.59 |

| | Unit | Hours | Hours £ | Mat'ls £ | O & P £ | Total £ |
|---|---|---|---|---|---|---|
| **Aluminium half round rainwater gutter with mastic joints, fixed to timber with support brackets at 1m maximum centres** | | | | | | |
| 100mm gutter | m | 0.30 | 4.35 | 12.29 | 2.50 | 19.14 |
| extra over for | | | | | | |
| running outlet | nr | 0.30 | 4.35 | 10.96 | 2.30 | 17.61 |
| angle | nr | 0.30 | 4.35 | 6.45 | 1.62 | 12.42 |
| stop end outlet | nr | 0.15 | 2.18 | 6.30 | 1.27 | 9.75 |
| stop end | nr | 0.15 | 2.18 | 2.43 | 0.69 | 5.30 |
| 125mm gutter | m | 0.32 | 4.64 | 15.49 | 3.02 | 23.15 |
| extra over for | | | | | | |
| running outlet | nr | 0.32 | 4.64 | 10.48 | 2.27 | 17.39 |
| angle | nr | 0.32 | 4.64 | 8.79 | 2.01 | 15.44 |
| stop end outlet | nr | 0.16 | 2.32 | 7.25 | 1.44 | 11.01 |
| stop end | nr | 0.16 | 2.32 | 3.16 | 0.82 | 6.30 |
| **Aluminium ogee rainwater gutter with mastic joints, fixed to timber with support brackets at 1m maximum centres** | | | | | | |
| 120 × 75mm gutter | m | 0.30 | 4.35 | 15.88 | 3.03 | 23.26 |
| extra over for | | | | | | |
| running outlet | nr | 0.30 | 4.35 | 20.00 | 3.65 | 28.00 |
| angle | nr | 0.30 | 4.35 | 16.75 | 3.17 | 24.27 |
| stop end outlet | nr | 0.15 | 2.18 | 27.95 | 4.52 | 34.64 |
| stop end | nr | 0.15 | 2.18 | 3.90 | 0.91 | 6.99 |
| 155 × 100mm gutter | m | 0.32 | 4.64 | 29.95 | 5.19 | 39.78 |
| extra over for | | | | | | |
| running outlet | nr | 0.32 | 4.64 | 26.45 | 4.66 | 35.75 |
| angle | nr | 0.32 | 4.64 | 25.86 | 4.58 | 35.08 |
| stop end outlet | nr | 0.16 | 2.32 | 21.55 | 3.58 | 27.45 |
| stop end | nr | 0.16 | 2.32 | 15.74 | 2.71 | 20.77 |

| | Unit | Hours | Hours £ | Mat'ls £ | O & P £ | Total £ |
|---|---|---|---|---|---|---|
| **Cast iron half-round rainwater gutter with mastic joints, primed, fixed to timber with support brackets at 1m maximum centres** | | | | | | |
| 100mm gutter | m | 0.36 | 5.22 | 10.85 | 2.41 | 18.48 |
| extra over for | | | | | | |
| running outlet | nr | 0.36 | 5.22 | 6.97 | 1.83 | 14.02 |
| angle | nr | 0.36 | 5.22 | 7.16 | 1.86 | 14.24 |
| stop end outlet | nr | 0.18 | 2.61 | 8.18 | 1.62 | 12.41 |
| stop end | nr | 0.18 | 2.61 | 2.41 | 0.75 | 5.77 |
| 115mm gutter | m | 0.38 | 5.51 | 11.20 | 2.51 | 19.22 |
| extra over for | | | | | | |
| running outlet | nr | 0.38 | 5.51 | 7.60 | 1.97 | 15.08 |
| angle | nr | 0.38 | 5.51 | 7.37 | 1.93 | 14.81 |
| stop end outlet | nr | 0.19 | 2.76 | 9.17 | 1.79 | 13.71 |
| stop end | nr | 0.19 | 2.76 | 3.12 | 0.88 | 6.76 |
| 125mm gutter | m | 0.40 | 5.80 | 12.95 | 2.81 | 21.56 |
| extra over for | | | | | | |
| running outlet | nr | 0.40 | 5.80 | 7.23 | 1.95 | 14.98 |
| angle | nr | 0.40 | 5.80 | 8.69 | 2.17 | 16.66 |
| stop end outlet | nr | 0.20 | 2.90 | 10.46 | 2.00 | 15.36 |
| stop end | nr | 0.20 | 2.90 | 3.12 | 0.90 | 6.92 |
| 150mm gutter | m | 0.44 | 6.38 | 21.64 | 4.20 | 32.22 |
| extra over for | | | | | | |
| running outlet | nr | 0.44 | 6.38 | 13.56 | 2.99 | 22.93 |
| angle | nr | 0.44 | 6.38 | 15.87 | 3.34 | 25.59 |
| stop end outlet | nr | 0.22 | 3.19 | 17.20 | 3.06 | 23.45 |
| stop end | nr | 0.22 | 3.19 | 4.32 | 1.13 | 8.64 |

| | Unit | Hours | Hours £ | Mat'ls £ | O & P £ | Total £ |
|---|---|---|---|---|---|---|
| **Roof outlets** | | | | | | |
| Cast iron circular roof outlet with flat grate, diameter | | | | | | |
| 50mm | nr | 0.50 | 7.25 | 74.61 | 12.28 | 94.14 |
| 75mm | nr | 0.60 | 8.70 | 81.34 | 13.51 | 103.55 |
| 100mm | nr | 0.70 | 10.15 | 95.34 | 15.82 | 121.31 |
| Aluminium circular roof outlet with domed grate, diameter | | | | | | |
| 50mm | nr | 0.50 | 7.25 | 68.85 | 11.42 | 87.52 |
| 75mm | nr | 0.60 | 8.70 | 78.60 | 13.10 | 100.40 |
| 100mm | nr | 0.70 | 10.15 | 101.46 | 16.74 | 128.35 |
| 150mm | nr | 0.80 | 11.60 | 113.55 | 18.77 | 143.92 |
| Plastic wire balloon guard for pipes and outlets, diameter | | | | | | |
| 50mm | nr | 0.05 | 0.73 | 1.98 | 0.41 | 3.11 |
| 63mm | nr | 0.05 | 0.73 | 2.10 | 0.42 | 3.25 |
| 75mm | nr | 0.05 | 0.73 | 2.16 | 0.43 | 3.32 |
| 100mm | nr | 0.05 | 0.73 | 2.18 | 0.44 | 3.34 |
| Galvanised wire balloon guard for pipes and outlets, diameter | | | | | | |
| 50mm | nr | 0.05 | 0.73 | 1.65 | 0.36 | 2.73 |
| 63mm | nr | 0.05 | 0.73 | 1.71 | 0.37 | 2.80 |
| 75mm | nr | 0.05 | 0.73 | 1.75 | 0.37 | 2.85 |
| 100mm | nr | 0.05 | 0.73 | 2.18 | 0.44 | 3.34 |
| Copper wire balloon guard for pipes and outlets, diameter | | | | | | |
| 50mm | nr | 0.05 | 0.73 | 1.97 | 0.40 | 3.10 |
| 63mm | nr | 0.05 | 0.73 | 2.00 | 0.41 | 3.13 |
| 75mm | nr | 0.05 | 0.73 | 2.15 | 0.43 | 3.31 |
| 100mm | nr | 0.05 | 0.73 | 3.02 | 0.56 | 4.31 |

| | Unit | Hours | Hours £ | Mat'ls £ | O & P £ | Total £ |
|---|---|---|---|---|---|---|
| **R11 DRAINAGE ABOVE GROUND** | | | | | | |
| **MuPVC waste system, solvent welded joints, clips at 500mm maximum centres, plugged to brickwork** | | | | | | |
| 32mm diameter pipe | m | 0.25 | 3.63 | 3.02 | 1.00 | 7.64 |
| extra over for | | | | | | |
| bend, 45 degrees | nr | 0.24 | 3.48 | 1.84 | 0.80 | 6.12 |
| bend, 87.5 degrees | nr | 0.24 | 3.48 | 1.84 | 0.80 | 6.12 |
| long tail bend, 90 degrees | nr | 0.24 | 3.48 | 1.84 | 0.80 | 6.12 |
| tee | nr | 0.24 | 3.48 | 3.14 | 0.99 | 7.61 |
| bottle trap | nr | 0.24 | 3.48 | 4.27 | 1.16 | 8.91 |
| tubular P trap | nr | 0.24 | 3.48 | 4.20 | 1.15 | 8.83 |
| tubular S trap | nr | 0.24 | 3.48 | 5.03 | 1.28 | 9.79 |
| 40mm diameter pipe | m | 0.28 | 4.06 | 3.25 | 1.10 | 8.41 |
| extra over for | | | | | | |
| bend, 45 degrees | nr | 0.26 | 3.77 | 2.11 | 0.88 | 6.76 |
| bend, 87.5 degrees | nr | 0.26 | 3.77 | 2.11 | 0.88 | 6.76 |
| long tail bend, 90 degrees | nr | 0.26 | 3.77 | 2.11 | 0.88 | 6.76 |
| tee | nr | 0.26 | 3.77 | 3.78 | 1.13 | 8.68 |
| cross tee | nr | 0.26 | 3.77 | 10.12 | 2.08 | 15.97 |
| bottle trap | nr | 0.26 | 3.77 | 4.85 | 1.29 | 9.91 |
| tubular P trap | nr | 0.26 | 3.77 | 5.03 | 1.32 | 10.12 |
| tubular S trap | nr | 0.26 | 3.77 | 5.99 | 1.46 | 11.22 |
| running P trap | nr | 0.26 | 3.77 | 7.20 | 1.65 | 12.62 |
| 50mm diameter pipe | m | 0.34 | 4.93 | 4.57 | 1.43 | 10.93 |
| extra over for | | | | | | |
| bend, 45 degrees | nr | 0.32 | 4.64 | 2.95 | 1.14 | 8.73 |
| bend, 87.5 degrees | nr | 0.32 | 4.64 | 2.95 | 1.14 | 8.73 |
| long tail bend, 90 degrees | nr | 0.32 | 4.64 | 2.95 | 1.14 | 8.73 |
| tee | nr | 0.32 | 4.64 | 6.34 | 1.65 | 12.63 |
| cross tee | nr | 0.32 | 4.64 | 11.74 | 2.46 | 18.84 |

| | Unit | Hours | Hours £ | Mat'ls £ | O & P £ | Total £ |
|---|---|---|---|---|---|---|
| **MuPVC waste system, push-fit joints, clips at 500mm maximum centres, plugged to brickwork** | | | | | | |
| 32mm diameter pipe | m | 0.20 | 2.90 | 1.09 | 0.60 | 4.59 |
| extra over for | | | | | | |
| bend, 45 degrees | nr | 0.18 | 2.61 | 0.92 | 0.53 | 4.06 |
| knuckle bend, 90 degrees | nr | 0.18 | 2.61 | 0.90 | 0.53 | 4.04 |
| spigot bend, 90 degrees | nr | 0.18 | 2.61 | 1.13 | 0.56 | 4.30 |
| tee | nr | 0.18 | 2.61 | 1.44 | 0.61 | 4.66 |
| P trap, 38mm | nr | 0.18 | 2.61 | 4.20 | 1.02 | 7.83 |
| bottle trap, 38mm | nr | 0.18 | 2.61 | 4.27 | 1.03 | 7.91 |
| anti-syphon bottle trap, 76mm | nr | 0.18 | 2.61 | 6.82 | 1.41 | 10.84 |
| connection to back inlet gulley, caulking bush | nr | 0.16 | 2.32 | 2.87 | 0.78 | 5.97 |
| 40mm diameter pipe | m | 0.24 | 3.48 | 1.84 | 0.80 | 6.12 |
| extra over for | | | | | | |
| bend, 45 degrees | nr | 0.20 | 2.90 | 0.98 | 0.58 | 4.46 |
| knuckle bend, 90 degrees | nr | 0.20 | 2.90 | 0.92 | 0.57 | 4.39 |
| spigot bend, 90 degrees | nr | 0.20 | 2.90 | 1.19 | 0.61 | 4.70 |
| tee | nr | 0.20 | 2.90 | 1.48 | 0.66 | 5.04 |
| P trap, 38mm | nr | 0.20 | 2.90 | 5.03 | 1.19 | 9.12 |
| bottle trap, 38mm | nr | 0.20 | 2.90 | 4.85 | 1.16 | 8.91 |
| anti-syphon bottle trap, 76mm | nr | 0.20 | 2.90 | 7.89 | 1.62 | 12.41 |
| connection to back inlet gulley, caulking bush | nr | 0.18 | 2.61 | 2.11 | 0.71 | 5.43 |
| **PVC-U soil system, solvent-welded joints, holderbats at 1250mm maximum centres, plugged to brickwork** | | | | | | |
| 82mm diameter pipe | m | 0.34 | 4.93 | 12.38 | 2.60 | 19.91 |
| extra over for | | | | | | |
| bend, 92.5 degrees | nr | 0.30 | 4.35 | 12.48 | 2.52 | 19.35 |
| bend, 135 degrees | nr | 0.30 | 4.35 | 12.48 | 2.52 | 19.35 |
| bend, spigot/spigot | nr | 0.30 | 4.35 | 12.48 | 2.52 | 19.35 |

| | Unit | Hours | Hours £ | Mat'ls £ | O & P £ | Total £ |
|---|---|---|---|---|---|---|
| **82mm diameter pipe (cont'd)** | | | | | | |
| branch, single, 92.5 degrees | nr | 0.38 | 5.51 | 18.35 | 3.58 | 27.44 |
| branch, single, 135 degrees | nr | 0.38 | 5.51 | 18.35 | 3.58 | 27.44 |
| access pipe, single branch | nr | 0.38 | 5.51 | 18.35 | 3.58 | 27.44 |
| access cap | nr | 0.22 | 3.19 | 11.26 | 2.17 | 16.62 |
| vent cowl | nr | 0.22 | 3.19 | 3.38 | 0.99 | 7.56 |
| 110mm diameter pipe | m | 0.38 | 5.51 | 11.53 | 2.56 | 19.60 |
| extra over for | | | | | | |
| bend, 92.5 degrees | nr | 0.34 | 4.93 | 14.82 | 2.96 | 22.71 |
| bend, 135 degrees | nr | 0.34 | 4.93 | 14.82 | 2.96 | 22.71 |
| bend, variable | nr | 0.34 | 4.93 | 14.82 | 2.96 | 22.71 |
| bend, spigot/spigot | nr | 0.34 | 4.93 | 14.82 | 2.96 | 22.71 |
| branch, single, 92.5 degrees | nr | 0.40 | 5.80 | 19.31 | 3.77 | 28.88 |
| branch, single, 135 degrees | nr | 0.40 | 5.80 | 19.31 | 3.77 | 28.88 |
| branch, single, 92.5 degrees spigot outlet | nr | 0.40 | 5.80 | 19.31 | 3.77 | 28.88 |
| access pipe connector | nr | 0.34 | 4.93 | 19.80 | 3.71 | 28.44 |
| access pipe, single branch | nr | 0.34 | 4.93 | 19.31 | 3.64 | 27.88 |
| access cap | nr | 0.38 | 5.51 | 11.89 | 2.61 | 20.01 |
| WC manifold connector | nr | 0.38 | 5.51 | 9.07 | 2.19 | 16.77 |
| vent cowl | nr | 0.22 | 3.19 | 3.56 | 1.01 | 7.76 |
| 150mm diameter pipe | m | 0.42 | 6.09 | 28.83 | 5.24 | 40.16 |
| extra over for | | | | | | |
| bend, 92.5 degrees | nr | 0.38 | 5.51 | 40.26 | 6.87 | 52.64 |
| bend, 135 degrees | nr | 0.38 | 5.51 | 40.26 | 6.87 | 52.64 |
| bend, spigot/spigot | nr | 0.38 | 5.51 | 40.26 | 6.87 | 52.64 |
| bend, spigot/socket | nr | 0.38 | 5.51 | 40.26 | 6.87 | 52.64 |
| branch, single, 92.5 degrees | nr | 0.42 | 6.09 | 66.31 | 10.86 | 83.26 |
| branch, single, 135 degrees | nr | 0.42 | 6.09 | 66.31 | 10.86 | 83.26 |
| access cap | nr | 0.40 | 5.80 | 32.62 | 5.76 | 44.18 |
| vent cowl | nr | 0.24 | 3.48 | 10.04 | 2.03 | 15.55 |

| | Unit | Hours | Hours £ | Mat'ls £ | O & P £ | Total £ |
|---|---|---|---|---|---|---|
| **Cast iron soil system with flexible joints, pipe brackets at 2m maximum centres, plugged to brickwork** | | | | | | |
| 50mm diameter pipe | m | 0.55 | 7.98 | 19.88 | 4.18 | 32.03 |
| extra over for | | | | | | |
| bend, short radius | nr | 0.45 | 6.53 | 9.84 | 2.45 | 18.82 |
| bend, short radius with door | nr | 0.45 | 6.53 | 24.26 | 4.62 | 35.40 |
| branch, single, plain | nr | 0.50 | 7.25 | 14.81 | 3.31 | 25.37 |
| branch, single with door | nr | 0.50 | 7.25 | 29.23 | 5.47 | 41.95 |
| branch, double, plain | nr | 0.50 | 7.25 | 26.05 | 5.00 | 38.30 |
| P trap with door | nr | 0.45 | 6.53 | 31.81 | 5.75 | 44.09 |
| 75mm diameter pipe | m | 0.60 | 8.70 | 19.67 | 4.26 | 32.63 |
| extra over for | | | | | | |
| bend, short radius | nr | 0.50 | 7.25 | 9.84 | 2.56 | 19.65 |
| bend, short radius with door | nr | 0.50 | 7.25 | 24.26 | 4.73 | 36.24 |
| bend, long radius | nr | 0.50 | 7.25 | 18.59 | 3.88 | 29.72 |
| bend, long radius, access | nr | 0.50 | 7.25 | 33.00 | 6.04 | 46.29 |
| branch, single, plain | nr | 0.55 | 7.98 | 14.81 | 3.42 | 26.20 |
| branch, single with door | nr | 0.55 | 7.98 | 29.23 | 5.58 | 42.79 |
| branch, double, plain | nr | 0.55 | 7.98 | 24.91 | 4.93 | 37.82 |
| offset, 75mm projection | nr | 0.50 | 7.25 | 9.72 | 2.55 | 19.52 |
| offset, 115mm projection | nr | 0.50 | 7.25 | 12.11 | 2.90 | 22.26 |
| offset, 150mm projection | nr | 0.50 | 7.25 | 12.11 | 2.90 | 22.26 |
| offset, 225mm projection | nr | 0.50 | 7.25 | 15.17 | 3.36 | 25.78 |
| offset, 300mm projection | nr | 0.50 | 7.25 | 17.89 | 3.77 | 28.91 |
| blank end, plain | nr | 0.50 | 7.25 | 3.54 | 1.62 | 12.41 |
| blank end, drilled and tapped | nr | 0.50 | 7.25 | 8.40 | 2.35 | 18.00 |
| P trap, plain | nr | 0.50 | 7.25 | 18.98 | 3.93 | 30.16 |
| P trap with door | nr | 0.50 | 7.25 | 33.40 | 6.10 | 46.75 |
| WC connector, 305mm effective length | nr | 0.50 | 7.25 | 18.17 | 3.81 | 29.23 |

| | Unit | Hours | Hours £ | Mat'ls £ | O & P £ | Total £ |
|---|---|---|---|---|---|---|
| 100mm diameter pipe | m | 0.65 | 9.43 | 29.47 | 5.83 | 44.73 |
| extra over for | | | | | | |
| bend, short radius | nr | 0.55 | 7.98 | 13.62 | 3.24 | 24.83 |
| bend, short radius with door | nr | 0.55 | 7.98 | 28.80 | 5.52 | 42.29 |
| bend, long radius | nr | 0.55 | 7.98 | 22.06 | 4.51 | 34.54 |
| bend, long radius, access | nr | 0.55 | 7.98 | 37.26 | 6.79 | 52.02 |
| bend, long radius, heel rest | nr | 0.55 | 7.98 | 26.71 | 5.20 | 39.89 |
| bend, long tail, 87.5 degrees | nr | 0.55 | 7.98 | 17.60 | 3.84 | 29.41 |
| branch, single, plain | nr | 0.60 | 8.70 | 21.05 | 4.46 | 34.21 |
| branch, single with door | nr | 0.60 | 8.70 | 36.27 | 6.75 | 51.72 |
| branch, double, plain | nr | 0.60 | 8.70 | 26.05 | 5.21 | 39.96 |
| branch, double with door | nr | 0.60 | 8.70 | 41.24 | 7.49 | 57.43 |
| branch, corner | nr | 0.60 | 8.70 | 34.27 | 6.45 | 49.42 |
| offset, 75mm projection | nr | 0.55 | 7.98 | 14.33 | 3.35 | 25.65 |
| offset, 115mm projection | nr | 0.55 | 7.98 | 17.08 | 3.76 | 28.81 |
| offset, 150mm projection | nr | 0.55 | 7.98 | 17.08 | 3.76 | 28.81 |
| offset, 225mm projection | nr | 0.55 | 7.98 | 19.58 | 4.13 | 31.69 |
| offset, 300mm projection | nr | 0.55 | 7.98 | 22.06 | 4.51 | 34.54 |
| blank end, plain | nr | 0.55 | 7.98 | 4.15 | 1.82 | 13.94 |
| blank end, drilled and tapped | nr | 0.55 | 7.98 | 9.02 | 2.55 | 19.54 |
| P trap, plain | nr | 0.55 | 7.98 | 21.83 | 4.47 | 34.28 |
| P trap with door | nr | 0.55 | 7.98 | 37.03 | 6.75 | 51.76 |
| WC connector, 305mm effective length | nr | 0.55 | 7.98 | 12.39 | 3.05 | 23.42 |
| 150mm diameter pipe | m | 0.70 | 10.15 | 57.23 | 10.11 | 77.49 |
| extra over for | | | | | | |
| bend, short radius | nr | 0.60 | 8.70 | 24.33 | 4.95 | 37.98 |
| bend, short radius with door | nr | 0.60 | 8.70 | 40.92 | 7.44 | 57.06 |
| bend, long radius | nr | 0.60 | 8.70 | 39.94 | 7.30 | 55.94 |
| bend, long radius, access | nr | 0.60 | 8.70 | 5.65 | 2.15 | 16.50 |
| branch, single, plain | nr | 0.65 | 9.43 | 43.31 | 7.91 | 60.65 |
| branch, single with door | nr | 0.65 | 9.43 | 59.95 | 10.41 | 79.78 |
| branch, double, plain | nr | 0.65 | 9.43 | 64.34 | 11.06 | 84.83 |
| blank end, plain | nr | 0.60 | 8.70 | 5.99 | 2.20 | 16.89 |
| blank end, drilled and tapped | nr | 0.60 | 8.70 | 109.00 | 17.66 | 135.36 |
| P trap with door | nr | 0.60 | 8.70 | 64.48 | 10.98 | 84.16 |

| | Unit | Hours | Hours £ | Mat'ls £ | O & P £ | Total £ |
|---|---|---|---|---|---|---|
| **MuPVC overflow system, solvent-welded joints, clips at 500mm maximum centres, plugged to brickwork** | | | | | | |
| 19mm diameter pipe | m | 0.20 | 2.90 | 1.72 | 0.69 | 5.31 |
| extra over for | | | | | | |
| bend | nr | 0.18 | 2.61 | 1.26 | 0.58 | 4.45 |
| tee | nr | 0.18 | 2.61 | 1.31 | 0.59 | 4.51 |
| connector, bent tank | nr | 0.20 | 2.90 | 1.87 | 0.72 | 5.49 |
| **Polypropylene traps, screwed joints to outlet and pipe** | | | | | | |
| Bottle trap, 38mm seal | | | | | | |
| 32mm | nr | 0.30 | 4.35 | 4.27 | 1.29 | 9.91 |
| 40mm | nr | 0.35 | 5.08 | 4.35 | 1.41 | 10.84 |
| Bottle trap, 76mm seal | | | | | | |
| 32mm | nr | 0.30 | 4.35 | 4.27 | 1.29 | 9.91 |
| 40mm | nr | 0.35 | 5.08 | 4.35 | 1.41 | 10.84 |
| Anti-syphon bottle trap | | | | | | |
| 32mm | nr | 0.30 | 4.35 | 6.82 | 1.68 | 12.85 |
| 40mm | nr | 0.35 | 5.08 | 7.09 | 1.82 | 13.99 |
| Tubular S trap | | | | | | |
| 32mm | nr | 0.30 | 4.35 | 5.03 | 1.41 | 10.79 |
| 40mm | nr | 0.35 | 5.08 | 5.99 | 1.66 | 12.72 |
| Tubular P trap | | | | | | |
| 32mm | nr | 0.30 | 4.35 | 4.20 | 1.28 | 9.83 |
| 40mm | nr | 0.35 | 5.08 | 5.03 | 1.52 | 11.62 |
| Running P trap | | | | | | |
| 40mm | nr | 0.35 | 5.08 | 7.20 | 1.84 | 14.12 |
| Bath trap | | | | | | |
| 40mm | nr | 0.35 | 5.08 | 6.32 | 1.71 | 13.10 |

| | Unit | Hours | Hours £ | Mat'ls £ | O & P £ | Total £ |
|---|---|---|---|---|---|---|
| Bath trap with overflow | | | | | | |
| 40mm | nr | 0.35 | 5.08 | 5.90 | 1.65 | 12.62 |
| Washing machine, half trap | | | | | | |
| 40mm | nr | 0.35 | 5.08 | 4.20 | 1.39 | 10.67 |

| | Unit | Hours | Hours £ | Mat'ls £ | O & P £ | Total £ |
|---|---|---|---|---|---|---|
| **R12 DRAINAGE BELOW GROUND** | | | | | | |
| **Timesaver cast iron drainage, jointed with flexible couplings and laid in trenches prepared by others** | | | | | | |
| 100mm diameter pipe | | | | | | |
| laid straight | m | 0.40 | 5.80 | 30.28 | 5.41 | 41.49 |
| laid vertically | m | 0.75 | 10.88 | 30.28 | 6.17 | 47.33 |
| in lengths less than 3m | m | 0.70 | 10.15 | 33.09 | 6.49 | 49.73 |
| Extra over for | | | | | | |
| standard coupling | nr | 0.15 | 2.18 | 18.23 | 3.06 | 23.47 |
| continuity clip for standard coupling | nr | 0.10 | 1.45 | 1.02 | 0.37 | 2.84 |
| stepped coupling | nr | 0.15 | 2.18 | 18.28 | 3.07 | 23.52 |
| continuity clip for stepped coupling | nr | 0.10 | 1.45 | 1.12 | 0.39 | 2.96 |
| hanging bracket | nr | 0.50 | 7.25 | 11.96 | 2.88 | 22.09 |
| bend, 87.5 degrees | nr | 0.40 | 5.80 | 36.09 | 6.28 | 48.17 |
| bend, 80 degrees | nr | 0.40 | 5.80 | 36.09 | 6.28 | 48.17 |
| bend, 67 degrees | nr | 0.40 | 5.80 | 35.10 | 6.14 | 47.04 |
| bend, 60 degrees | nr | 0.40 | 5.80 | 35.10 | 6.14 | 47.04 |
| bend, 45 degrees | nr | 0.40 | 5.80 | 33.71 | 5.93 | 45.44 |
| bend, 35 degrees | nr | 0.40 | 5.80 | 33.71 | 5.93 | 45.44 |
| bend, 22.5 degrees | nr | 0.40 | 5.80 | 31.53 | 5.60 | 42.93 |
| bend, 10 degrees | nr | 0.40 | 5.80 | 29.25 | 5.26 | 40.31 |
| bend, 87.5 degrees, heel rest | nr | 0.40 | 5.80 | 32.75 | 5.78 | 44.33 |
| bend, access rear, 87.5 degrees | nr | 0.40 | 5.80 | 68.36 | 11.12 | 85.28 |
| bend, access rear, 45 degrees | nr | 0.40 | 5.80 | 68.36 | 11.12 | 85.28 |
| bend, access side, 87.5 degrees | nr | 0.40 | 5.80 | 74.34 | 12.02 | 92.16 |
| bend, access side, 45 degrees | nr | 0.40 | 5.80 | 74.34 | 12.02 | 92.16 |
| bend, long radius, 87.5 degrees | nr | 0.40 | 5.80 | 49.27 | 8.26 | 63.33 |
| bend, long radius, 87.5 degrees, with heel rest | nr | 0.40 | 5.80 | 51.20 | 8.55 | 65.55 |
| bend, long tail, 87.5 degrees | nr | 0.40 | 5.80 | 67.01 | 10.92 | 83.73 |

| | Unit | Hours | Hours £ | Mat'ls £ | O & P £ | Total £ |
|---|---|---|---|---|---|---|
| bend, large radius, 87.5 degrees | nr | 0.40 | 5.80 | 71.53 | 11.60 | 88.93 |
| branch, 100 × 100mm, 87.5 degrees | nr | 0.60 | 8.70 | 47.50 | 8.43 | 64.63 |
| branch, 100 × 100mm, 67.5 degrees | nr | 0.60 | 8.70 | 49.03 | 8.66 | 66.39 |
| branch, 100 × 100mm, 45 degrees | nr | 0.60 | 8.70 | 48.52 | 8.58 | 65.80 |
| branch, 100 × 100mm, 87.5 degrees, right-hand access | nr | 0.60 | 8.70 | 81.22 | 13.49 | 103.41 |
| branch, 100 × 100mm, 45 degrees, right-hand access | nr | 0.60 | 8.70 | 81.22 | 13.49 | 103.41 |
| branch, 100 × 100mm, 87.5 degrees, rear access | nr | 0.60 | 8.70 | 81.22 | 13.49 | 103.41 |
| branch, 100 × 100mm, 45 degrees, rear access | nr | 0.60 | 8.70 | 81.22 | 13.49 | 103.41 |
| branch, 100 × 100mm, 87.5 degrees, left-hand access | nr | 0.60 | 8.70 | 81.22 | 13.49 | 103.41 |
| branch, 100 × 100mm, 45 degrees, left-hand access | nr | 0.60 | 8.70 | 81.22 | 13.49 | 103.41 |
| branch, 100 × 100mm, 87.5 degrees, double plain | nr | 0.60 | 8.70 | 63.02 | 10.76 | 82.48 |
| socket ferrule with cap | nr | 0.40 | 5.80 | 31.63 | 5.61 | 43.04 |
| blank cap plain | nr | 0.40 | 5.80 | 14.07 | 2.98 | 22.85 |
| cleaning arm bend | nr | 0.60 | 8.70 | 51.12 | 8.97 | 68.79 |
| transitional pipe, socket for clayware | nr | 0.40 | 5.80 | 36.85 | 6.40 | 49.05 |
| transitional pipe, socket for cast iron | nr | 0.40 | 5.80 | 28.69 | 5.17 | 39.66 |
| 150mm diameter pipe | | | | | | |
| laid straight | m | 0.60 | 8.70 | 54.23 | 9.44 | 72.37 |
| laid vertically | m | 0.80 | 11.60 | 54.23 | 9.87 | 75.70 |
| in lengths less than 3m | m | 0.90 | 13.05 | 56.23 | 10.39 | 79.67 |
| Extra over for | | | | | | |
| standard coupling | nr | 0.15 | 2.18 | 21.21 | 3.51 | 26.89 |
| continuity clip for standard coupling | nr | 0.10 | 1.45 | 1.21 | 0.40 | 3.06 |

| | Unit | Hours | Hours £ | Mat'ls £ | O & P £ | Total £ |
|---|---|---|---|---|---|---|
| stepped coupling | nr | 0.15 | 2.18 | 1.32 | 0.52 | 4.02 |
| continuity clip for stepped coupling | nr | 0.10 | 1.45 | 1.37 | 0.42 | 3.24 |
| hanging bracket | nr | 0.50 | 7.25 | 14.77 | 3.30 | 25.32 |
| bend, 87.5 degrees | nr | 0.70 | 10.15 | 71.03 | 12.18 | 93.36 |
| bend, 67.5 degrees | nr | 0.70 | 10.15 | 62.56 | 10.91 | 83.62 |
| bend, 45 degrees | nr | 0.70 | 10.15 | 53.59 | 9.56 | 73.30 |
| bend, 35 degrees | nr | 0.70 | 10.15 | 53.59 | 9.56 | 73.30 |
| bend, 22.5 degrees | nr | 0.70 | 10.15 | 51.43 | 9.24 | 70.82 |
| bend, 10 degrees | nr | 0.70 | 10.15 | 47.76 | 8.69 | 66.60 |
| bend, 87.5 degrees, heel rest | nr | 0.70 | 10.15 | 79.32 | 13.42 | 102.89 |
| bend, access rear, 87.5 degrees | nr | 0.70 | 10.15 | 138.35 | 22.28 | 170.78 |
| bend, access rear, 45 degrees | nr | 0.70 | 10.15 | 138.35 | 22.28 | 170.78 |
| bend, access side, 87.5 degrees | nr | 0.70 | 10.15 | 138.35 | 22.28 | 170.78 |
| bend, access side, 45 degrees | nr | 0.70 | 10.15 | 138.35 | 22.28 | 170.78 |
| bend, long radius, 87.5 degrees | nr | 0.70 | 10.15 | 48.18 | 8.75 | 67.08 |
| bend, long radius, 87.5 degrees, with heel rest | nr | 0.70 | 10.15 | 84.04 | 14.13 | 108.32 |
| branch, 150 × 100mm, 87.5 degrees | nr | 0.90 | 13.05 | 89.63 | 15.40 | 118.08 |
| branch, 150 × 100mm, 67.5 degrees | nr | 0.90 | 13.05 | 91.49 | 15.68 | 120.22 |
| branch, 150 × 100mm, 45 degrees | nr | 0.90 | 13.05 | 86.86 | 14.99 | 114.90 |
| branch, 150 × 100mm, 87.5 degrees, right-hand access | nr | 0.90 | 13.05 | 87.22 | 15.04 | 115.31 |
| branch, 150 × 100mm, 45 degrees, right-hand access | nr | 0.90 | 13.05 | 87.22 | 15.04 | 115.31 |
| branch, 150 × 100mm, 87.5 degrees, rear access | nr | 0.90 | 13.05 | 169.17 | 27.33 | 209.55 |
| branch, 150 × 100mm, 45 degrees, rear access | nr | 0.90 | 13.05 | 169.71 | 27.41 | 210.17 |
| branch, 150 × 100mm, 87.5 degrees, left hand access | nr | 0.90 | 13.05 | 169.52 | 27.39 | 209.96 |
| socket ferrule with cap | nr | 0.50 | 7.25 | 57.08 | 9.65 | 73.98 |
| blank cap plain | nr | 0.50 | 7.25 | 27.24 | 5.17 | 39.66 |
| transitional pipe, socket for clayware | nr | 0.60 | 8.70 | 48.64 | 8.60 | 65.94 |

| | Unit | Hours | Hours £ | Mat'ls £ | O & P £ | Total £ |
|---|---|---|---|---|---|---|
| transitional pipe, socket for cast iron | nr | 0.60 | 8.70 | 47.17 | 8.38 | 64.25 |
| gulley trap plain, 150mm × 87.5 degrees | nr | 0.90 | 13.05 | 96.32 | 16.41 | 125.78 |
| 225mm diameter pipe | | | | | | |
| laid straight | m | 1.10 | 15.95 | 137.93 | 23.08 | 176.96 |
| laid vertically | m | 1.30 | 18.85 | 137.93 | 23.52 | 180.30 |
| in lengths less than 3m | m | 1.50 | 21.75 | 142.04 | 24.57 | 188.36 |
| Extra over for | | | | | | |
| standard coupling | nr | 0.20 | 2.90 | 1.81 | 0.71 | 5.42 |
| continuity clip for standard coupling | nr | 0.15 | 2.18 | 1.62 | 0.57 | 4.36 |
| bend, 87.5 degrees | nr | 1.00 | 14.50 | 191.86 | 30.95 | 237.31 |
| bend, 45 degrees | nr | 1.00 | 14.50 | 191.86 | 30.95 | 237.31 |
| bend, 22.5 degrees | nr | 1.00 | 14.50 | 191.86 | 30.95 | 237.31 |
| bend, 10 degrees | nr | 1.00 | 14.50 | 191.86 | 30.95 | 237.31 |
| bend, access rear, 87.5 degrees | nr | 1.00 | 14.50 | 291.73 | 45.93 | 352.16 |
| bend, access rear, 45 degrees | nr | 1.00 | 14.50 | 291.73 | 45.93 | 352.16 |
| bend, access side, 87.5 degrees | nr | 1.00 | 14.50 | 282.98 | 44.62 | 342.10 |
| bend, access side, 45 degrees | nr | 1.00 | 14.50 | 282.98 | 44.62 | 342.10 |
| branch, 225 × 100mm, 87.5 degrees | nr | 1.20 | 17.40 | 239.91 | 38.60 | 295.91 |
| branch, 225 × 100mm, 45 degrees | nr | 1.20 | 17.40 | 239.91 | 38.60 | 295.91 |
| branch, 225 × 150mm, 87.5 degrees | nr | 1.20 | 17.40 | 251.12 | 40.28 | 308.80 |
| branch, 225 × 150mm, 45 degrees | nr | 1.20 | 17.40 | 251.12 | 40.28 | 308.80 |
| branch, 225 × 225mm, 87.5 degrees | nr | 1.20 | 17.40 | 272.65 | 43.51 | 333.56 |
| branch, 225 × 225mm, 45 degrees | nr | 1.20 | 17.40 | 272.65 | 43.51 | 333.56 |
| branch, 225 × 100mm, 45 degrees, right-hand access | nr | 1.20 | 17.40 | 349.43 | 55.02 | 421.85 |
| branch, 225 × 150mm, 87.5 degrees, right-hand access | nr | 1.20 | 17.40 | 359.74 | 56.57 | 433.71 |

| | Unit | Hours | Hours £ | Mat'ls £ | O & P £ | Total £ |
|---|---|---|---|---|---|---|
| branch, 225 × 150mm, 45 degrees, right-hand access | nr | 1.20 | 17.40 | 359.74 | 56.57 | 433.71 |
| branch, 225 × 225mm, 87.5 degrees, right-hand access | nr | 1.20 | 17.40 | 372.02 | 58.41 | 447.83 |
| branch, 225 × 225mm, 45 degrees, right-hand access | nr | 1.20 | 17.40 | 372.02 | 58.41 | 447.83 |
| branch, 225 × 100mm, 45 degrees, rear access | nr | 1.20 | 17.40 | 343.43 | 54.12 | 414.95 |
| branch, 225 × 150mm, 87.5 degrees, rear access | nr | 1.20 | 17.40 | 343.43 | 54.12 | 414.95 |
| branch, 225 × 150mm, 45 degrees, rear access | nr | 1.20 | 17.40 | 359.74 | 56.57 | 433.71 |
| branch, 225 × 225mm, 87.5 degrees, rear access | nr | 1.20 | 17.40 | 372.02 | 58.41 | 447.83 |
| branch, 225 × 225mm, 45 degrees, rear access | nr | 1.20 | 17.40 | 372.02 | 58.41 | 447.83 |
| branch, 225 × 100mm, 87.5 degrees, left-hand access | nr | 1.20 | 17.40 | 343.43 | 54.12 | 414.95 |
| branch, 225 × 100mm, 45 degrees, left-hand access | nr | 1.20 | 17.40 | 343.43 | 54.12 | 414.95 |
| branch, 225 × 150mm, 87.5 degrees, left-hand access | nr | 1.20 | 17.40 | 359.74 | 56.57 | 433.71 |
| branch, 225 × 150mm, 45 degrees, left-hand access | nr | 1.20 | 17.40 | 359.74 | 56.57 | 433.71 |
| branch, 225 × 225mm, 87.5 degrees, left-hand access | nr | 1.20 | 17.40 | 372.02 | 58.41 | 447.83 |
| branch, 225 × 225mm, 45 degrees, left-hand access | nr | 1.20 | 17.40 | 372.02 | 58.41 | 447.83 |
| diminishing pipe 225-100mm | nr | 0.70 | 10.15 | 91.32 | 15.22 | 116.69 |
| diminishing pipe 225-150mm | nr | 0.70 | 10.15 | 91.32 | 15.22 | 116.69 |
| socket ferrule with cap | nr | 0.50 | 7.25 | 217.49 | 33.71 | 258.45 |
| blank cap plain | nr | 0.50 | 7.25 | 80.99 | 13.24 | 101.48 |
| transitional pipe, socket for clayware | nr | 0.60 | 8.70 | 137.71 | 21.96 | 168.37 |
| transitional pipe, socket for cast iron | nr | 0.60 | 8.70 | 145.79 | 23.17 | 177.66 |

| | Unit | Hours | Hours £ | Mat'ls £ | O & P £ | Total £ |
|---|---|---|---|---|---|---|
| **Gullies and traps** | | | | | | |
| Gulley trap plain, 100mm × 87.5 degrees | nr | 0.70 | 10.15 | 49.50 | 8.95 | 68.60 |
| Gulley trap plain, 100mm × 45 degrees | nr | 0.70 | 10.15 | 55.16 | 9.80 | 75.11 |
| Gulley trap plain, 150mm × 87.5 degrees | nr | 0.70 | 10.15 | 96.32 | 15.97 | 122.44 |
| Gulley trap, 100mm × 87.5 degrees, surface access | nr | 0.70 | 10.15 | 91.13 | 15.19 | 116.47 |
| Gulley trap, 100mm × 87.5 degrees, 225mm inlet | nr | 0.70 | 10.15 | 76.15 | 12.95 | 99.25 |
| Gulley trap running, 150mm double access, 100 × 87.5 degrees | nr | 0.90 | 13.05 | 191.86 | 30.74 | 235.65 |
| Gulley trap running, 150mm double access, 150 × 87.5 degrees | nr | 0.90 | 13.05 | 302.48 | 47.33 | 362.86 |
| Gulley trap running, 150mm double access, 225 × 87.5 degrees | nr | 0.90 | 13.05 | 328.72 | 51.27 | 393.04 |
| Gulley inlet plain | nr | 0.90 | 13.05 | 47.24 | 9.04 | 69.33 |
| Gulley inlet, 100 × 100mm with horizontal branch | nr | 0.90 | 13.05 | 54.70 | 10.16 | 77.91 |
| Gulley inlet, 100 × 100mm with vertical branch | nr | 0.90 | 13.05 | 55.51 | 10.28 | 78.84 |
| Raising piece, plain, 100 × 75mm | nr | 0.40 | 5.80 | 38.62 | 6.66 | 51.08 |
| Raising piece, plain, 100 × 150mı | nr | 0.40 | 5.80 | 41.40 | 7.08 | 54.28 |
| Raising piece, plain, 100 × 225mı | nr | 0.40 | 5.80 | 43.59 | 7.41 | 56.80 |
| Raising piece, plain, 100 × 300mı | nr | 0.40 | 5.80 | 48.78 | 8.19 | 62.77 |
| Notched grate, 200mm | nr | 0.10 | 1.45 | 5.14 | 0.99 | 7.58 |
| Plain grate, 200mm | nr | 0.10 | 1.45 | 4.71 | 0.92 | 7.08 |
| Solid cover, 200mm | nr | 0.20 | 2.90 | 6.86 | 1.46 | 11.22 |
| Hinged and locking grate, 200mr | nr | 0.25 | 3.63 | 27.83 | 4.72 | 36.17 |
| Sealed plate and frame, 200mm | nr | 0.25 | 3.63 | 28.89 | 4.88 | 37.39 |
| Trapless gulley, 87.5 degrees × 100mm | nr | 2.00 | 29.00 | 26.30 | 8.30 | 63.60 |

| | Unit | Hours | Hours £ | Mat'ls £ | O & P £ | Total £ |
|---|---|---|---|---|---|---|
| Dean's trap, 87.5 degrees × 100mm with galvanised sediment pan | nr | 3.00 | 43.50 | 129.89 | 26.01 | 199.40 |
| Gulley trap 425mm deep, 87.5 degrees × 100mm with galvanised sediment pan | nr | 2.70 | 39.15 | 129.89 | 25.36 | 194.40 |
| Gulley trap with square top, 87.5 degrees × 100mm outlet with galvanised sediment pan and grate | nr | 2.70 | 39.15 | 198.71 | 35.68 | 273.54 |
| Garage gulley 300 × 600mm, 87.5 degrees × 100mm outlet with galvanised sediment pan and grate | nr | 2.80 | 40.60 | 392.04 | 64.90 | 497.54 |
| Garage gulley 300 × 600mm, 45 degrees × 100mm outlet with galvanised sediment pan and grate | nr | 2.80 | 40.60 | 392.04 | 64.90 | 497.54 |
| Rainwater shoe with 100mm horizontal outlet | nr | 0.80 | 11.60 | 78.98 | 13.59 | 104.17 |

## S10 HOT AND COLD WATER

**Copper pipe, lead-free pre-soldered capillary joints and fittings, clips at 1250mm maximum centres**

| | Unit | Hours | Hours £ | Mat'ls £ | O & P £ | Total £ |
|---|---|---|---|---|---|---|
| 8mm diameter, to timber | m | 0.20 | 2.90 | 1.21 | 0.62 | 4.73 |
| 8mm diameter, plugged and screwed | m | 0.22 | 3.19 | 1.49 | 0.70 | 5.38 |
| extra over for | | | | | | |
| straight coupling | nr | 0.18 | 2.61 | 1.01 | 0.54 | 4.16 |
| reduced coupling, 8 × 6mm | nr | 0.18 | 2.61 | 3.65 | 0.94 | 7.20 |
| straight female connector, 8mm × 1/4in | nr | 0.18 | 2.61 | 3.86 | 0.97 | 7.44 |
| straight male connector, 8mm × 1/4in | nr | 0.18 | 2.61 | 2.59 | 0.78 | 5.98 |
| reducer, 8 × 6mm | nr | 0.18 | 2.61 | 2.91 | 0.83 | 6.35 |
| elbow, 8mm | nr | 0.18 | 2.61 | 2.19 | 0.72 | 5.52 |
| street elbow, 8mm | nr | 0.18 | 2.61 | 2.99 | 0.84 | 6.44 |
| equal tee, 8mm | nr | 0.22 | 3.19 | 3.50 | 1.00 | 7.69 |
| stop end, 8mm | nr | 0.18 | 2.61 | 3.17 | 0.87 | 6.65 |
| 10mm diameter, to timber | m | 0.20 | 2.90 | 1.42 | 0.65 | 4.97 |
| 10mm diameter, plugged and screwed | m | 0.22 | 3.19 | 1.69 | 0.73 | 5.61 |
| extra over for | | | | | | |
| straight coupling | nr | 0.18 | 2.61 | 0.52 | 0.47 | 3.60 |
| reduced coupling, 10 × 8mm | nr | 0.18 | 2.61 | 3.20 | 0.87 | 6.68 |
| reducer, 10 × 8mm | nr | 0.18 | 2.61 | 2.24 | 0.73 | 5.58 |
| elbow, 10mm | nr | 0.18 | 2.61 | 0.89 | 0.53 | 4.03 |
| street elbow, 10mm | nr | 0.18 | 2.61 | 2.60 | 0.78 | 5.99 |
| equal tee, 10mm | nr | 0.22 | 3.19 | 2.03 | 0.78 | 6.00 |
| elbow, 10mm | nr | 0.18 | 2.61 | 1.71 | 0.65 | 4.97 |
| made bend | nr | 0.18 | 2.61 | 0.00 | 0.39 | 3.00 |

| | Unit | Hours | Hours £ | Mat'ls £ | O & P £ | Total £ |
|---|---|---|---|---|---|---|
| 15mm diameter, to timber | m | 0.20 | 2.90 | 1.71 | 0.69 | 5.30 |
| 15mm diameter, plugged and screwed | m | 0.22 | 3.19 | 1.97 | 0.77 | 5.93 |
| extra over for | | | | | | |
| reduced coupling, 15 × 8mm | nr | 0.18 | 2.61 | 2.27 | 0.73 | 5.61 |
| reduced coupling, 15 × 10mm | nr | 0.18 | 2.61 | 2.21 | 0.72 | 5.54 |
| reduced coupling, 15 × 12mm | nr | 0.18 | 2.61 | 1.68 | 0.64 | 4.93 |
| adaptor coupling, 15mm × 1/2in | nr | 0.18 | 2.61 | 2.59 | 0.78 | 5.98 |
| straight female connector, 15mm × 1/2in | nr | 0.18 | 2.61 | 2.72 | 0.80 | 6.13 |
| female reducing connector, 15mm × 3/8in | nr | 0.18 | 2.61 | 5.42 | 1.20 | 9.23 |
| female reducing connector, 15mm × 1/4in | nr | 0.18 | 2.61 | 4.83 | 1.12 | 8.56 |
| straight male connector, 15mm × 1/2in | nr | 0.18 | 2.61 | 2.33 | 0.74 | 5.68 |
| male reducing connector, 15mm × 3/4in | nr | 0.18 | 2.61 | 3.16 | 0.87 | 6.64 |
| tank connector, 15mm × 1/2in | nr | 0.18 | 2.61 | 5.78 | 1.26 | 9.65 |
| reducer, 15 × 8mm | nr | 0.18 | 2.61 | 1.86 | 0.67 | 5.14 |
| reducer, 15 × 10mm | nr | 0.18 | 2.61 | 0.73 | 0.50 | 3.84 |
| reducer, 15 × 12mm | nr | 0.18 | 2.61 | 1.46 | 0.61 | 4.68 |
| female adaptor, 15 × 1/2in | nr | 0.18 | 2.61 | 4.68 | 1.09 | 8.38 |
| male adaptor, 15 × 1/2in | nr | 0.18 | 2.61 | 4.78 | 1.11 | 8.50 |
| adaptor, 15 × 1/2in | nr | 0.18 | 2.61 | 2.27 | 0.73 | 5.61 |
| male elbow, 15mm × 1/2in | nr | 0.18 | 2.61 | 4.63 | 1.09 | 8.33 |
| reducing male elbow, 15mm × 3/8in | nr | 0.18 | 2.61 | 6.52 | 1.37 | 10.50 |
| female elbow, 15mm × 1/2in | nr | 0.18 | 2.61 | 4.00 | 0.99 | 7.60 |
| reducing female elbow, 15mm × 1/4in | nr | 0.18 | 2.61 | 6.38 | 1.35 | 10.34 |
| backplate elbow, 15mm × 1/2in | nr | 0.18 | 2.61 | 10.29 | 1.94 | 14.84 |
| flanged bend, 15mm × 1/2in | nr | 0.18 | 2.61 | 4.85 | 1.12 | 8.58 |
| flanged bend, 15mm × 3/4in | nr | 0.18 | 2.61 | 10.89 | 2.03 | 15.53 |
| slow bend | nr | 0.18 | 2.61 | 0.78 | 0.51 | 3.90 |
| return bend | nr | 0.18 | 2.61 | 7.27 | 1.48 | 11.36 |
| obtuse elbow | nr | 0.18 | 2.61 | 0.93 | 0.53 | 4.07 |

| | Unit | Hours | Hours £ | Mat'ls £ | O & P £ | Total £ |
|---|---|---|---|---|---|---|
| **Extra over 15mm diameter pipework (cont'd)** | | | | | | |
| tee, reduced branch largest end 15mm | nr | 0.22 | 3.19 | 8.77 | 1.79 | 13.75 |
| tee, both ends reduced, largest end 15mm | nr | 0.22 | 3.19 | 8.18 | 1.71 | 13.08 |
| sweep tee, 90 degrees, 15mm | nr | 0.22 | 3.19 | 8.20 | 1.71 | 13.10 |
| offset tee, 15mm | nr | 0.22 | 3.19 | 10.84 | 2.10 | 16.13 |
| double sweep tee, 15mm | nr | 0.22 | 3.19 | 9.27 | 1.87 | 14.33 |
| cross equal tee, 15mm | nr | 0.22 | 3.19 | 12.27 | 2.32 | 17.78 |
| stop end, 15mm | nr | 0.18 | 2.61 | 3.17 | 0.87 | 6.65 |
| tap connector, 15mm | nr | 0.18 | 2.61 | 3.61 | 0.93 | 7.15 |
| bent union adaptor, 15mm × 3/4in | nr | 0.18 | 2.61 | 8.03 | 1.60 | 12.24 |
| bent male union connector, 15mm × 1/2in | nr | 0.18 | 2.61 | 9.45 | 1.81 | 13.87 |
| bent female union connector, 15mm × 1/2in | nr | 0.18 | 2.61 | 9.45 | 1.81 | 13.87 |
| made bend | nr | 0.18 | 2.61 | 0.00 | 0.39 | 3.00 |
| 22mm diameter, to timber | m | 0.22 | 3.19 | 2.68 | 0.88 | 6.75 |
| 22mm diameter, plugged and screwed | m | 0.24 | 3.48 | 2.91 | 0.96 | 7.35 |
| extra over for | | | | | | |
| straight coupling | nr | 0.20 | 2.90 | 3.21 | 0.92 | 7.03 |
| reduced coupling, 22 × 8mm | nr | 0.20 | 2.90 | 3.21 | 0.92 | 7.03 |
| reduced coupling, 22 × 10mm | nr | 0.20 | 2.90 | 3.21 | 0.92 | 7.03 |
| reduced coupling, 22 × 15mm | nr | 0.20 | 2.90 | 2.16 | 0.76 | 5.82 |
| straight female connector, 22mm × 3/4in | nr | 0.20 | 2.90 | 3.94 | 1.03 | 7.87 |
| female reducing connector, 22mm × 1/2in | nr | 0.20 | 2.90 | 6.26 | 1.37 | 10.53 |
| straight male connector, 22mm × 3/4in | nr | 0.20 | 2.90 | 4.12 | 1.05 | 8.07 |
| male reducing connector, 22mm × 1/2in | nr | 0.20 | 2.90 | 6.48 | 1.41 | 10.79 |

| | Unit | Hours | Hours £ | Mat'ls £ | O & P £ | Total £ |
|---|---|---|---|---|---|---|
| **Extra over 22mm diameter pipework (cont'd)** | | | | | | |
| tank connector, 22mm × 1in | nr | 0.20 | 2.90 | 8.81 | 1.76 | 13.47 |
| reducer, 22 × 12mm | nr | 0.20 | 2.90 | 3.33 | 0.93 | 7.16 |
| female adaptor, 22 × 3/4in | nr | 0.20 | 2.90 | 8.81 | 1.76 | 13.47 |
| male adaptor, 22 × 3/4in | nr | 0.20 | 2.90 | 6.10 | 1.35 | 10.35 |
| adaptor, 22 × 3/4in | nr | 0.20 | 2.90 | 2.51 | 0.81 | 6.22 |
| street elbow, 22mm | nr | 0.20 | 2.90 | 1.98 | 0.73 | 5.61 |
| male elbow, 22mm × 3/4in | nr | 0.20 | 2.90 | 6.57 | 1.42 | 10.89 |
| female elbow, 22mm × 3/4in | nr | 0.20 | 2.90 | 5.78 | 1.30 | 9.98 |
| reducing female elbow, 22mm × 1/2in | nr | 0.20 | 2.90 | 6.68 | 1.44 | 11.02 |
| backplate elbow, 22mm × 3/4in | nr | 0.20 | 2.90 | 10.42 | 2.00 | 15.32 |
| flanged bend, 22mm × 3/4in | nr | 0.20 | 2.90 | 11.99 | 2.23 | 17.12 |
| overflow bend, 22mm × 3/4in | nr | 0.20 | 2.90 | 14.69 | 2.64 | 20.23 |
| slow bend | nr | 0.20 | 2.90 | 1.53 | 0.66 | 5.09 |
| return bend | nr | 0.20 | 2.90 | 14.29 | 2.58 | 19.77 |
| obtuse elbow | nr | 0.20 | 2.90 | 1.94 | 0.73 | 5.57 |
| tee, reduced branch largest end 22mm | nr | 0.24 | 3.48 | 6.39 | 1.48 | 11.35 |
| tee, both ends reduced, largest end 22mm | nr | 0.24 | 3.48 | 2.76 | 0.94 | 7.18 |
| corner tee, 22mm | nr | 0.24 | 3.48 | 13.31 | 2.52 | 19.31 |
| sweep tee, 90 degrees, 22mm | nr | 0.24 | 3.48 | 10.56 | 2.11 | 16.15 |
| offset tee, 22mm | nr | 0.24 | 3.48 | 13.05 | 2.48 | 19.01 |
| double sweep tee, 22mm | nr | 0.24 | 3.48 | 12.62 | 2.42 | 18.52 |
| cross equal tee, 22mm | nr | 0.24 | 3.48 | 13.71 | 2.58 | 19.77 |
| stop end, 22mm | nr | 0.20 | 2.90 | 2.23 | 0.77 | 5.90 |
| tap connector, 22mm × 1/2in | nr | 0.20 | 2.90 | 9.27 | 1.83 | 14.00 |
| bent tap connector, 22mm × 1/2in | nr | 0.20 | 2.90 | 11.30 | 2.13 | 16.33 |
| bent tap connector, 22mm × 3/4in | nr | 0.20 | 2.90 | 5.65 | 1.28 | 9.83 |
| bent union adaptor, 22mm × 1in | nr | 0.20 | 2.90 | 7.54 | 1.57 | 12.01 |
| bent male union connector, 22mm × 3/4in | nr | 0.20 | 2.90 | 9.28 | 1.83 | 14.01 |

| | Unit | Hours | Hours £ | Mat'ls £ | O & P £ | Total £ |
|---|---|---|---|---|---|---|
| **Extra over 22mm diameter pipework (cont'd)** | | | | | | |
| bent female union connector, 22mm × 3/4in | nr | 0.20 | 2.90 | 12.30 | 2.28 | 17.48 |
| made bend | nr | 0.20 | 2.90 | 0.00 | 0.44 | 3.34 |
| 28mm diameter, to timber | m | 0.23 | 3.34 | 3.80 | 1.07 | 8.21 |
| 28mm diameter, plugged and screwed | m | 0.25 | 3.63 | 4.05 | 1.15 | 8.83 |
| extra over for | | | | | | |
| straight coupling | nr | 0.24 | 3.48 | 1.11 | 0.69 | 5.28 |
| reduced coupling, 28 × 15mm | nr | 0.24 | 3.48 | 4.94 | 1.26 | 9.68 |
| reduced coupling, 28 × 22mm | nr | 0.24 | 3.48 | 3.02 | 0.98 | 7.48 |
| adaptor coupling, 28mm × 1in | nr | 0.24 | 3.48 | 4.48 | 1.19 | 9.15 |
| straight female connector, 28mm × 1in | nr | 0.24 | 3.48 | 7.42 | 1.64 | 12.54 |
| straight male connector, 28mm × 1in | nr | 0.24 | 3.48 | 6.65 | 1.52 | 11.65 |
| male reducing connector, 28mm × 3/4in | nr | 0.24 | 3.48 | 10.20 | 2.05 | 15.73 |
| tank connector, 28mm × 1in | nr | 0.24 | 3.48 | 11.58 | 2.26 | 17.32 |
| tank connector, 28mm × 1in | nr | 0.24 | 3.48 | 11.58 | 2.26 | 17.32 |
| reducer, 28 × 15mm | nr | 0.24 | 3.48 | 2.44 | 0.89 | 6.81 |
| reducer, 28 × 22mm | nr | 0.24 | 3.48 | 1.86 | 0.80 | 6.14 |
| female adaptor, 28mm × 1in | nr | 0.24 | 3.48 | 10.05 | 2.03 | 15.56 |
| male adaptor, 28mm × 1in | nr | 0.24 | 3.48 | 10.18 | 2.05 | 15.71 |
| adaptor, 28mm × 1in | nr | 0.24 | 3.48 | 3.56 | 1.06 | 8.10 |
| reducing elbow, 28 × 22mm | nr | 0.24 | 3.48 | 8.62 | 1.82 | 13.92 |
| street elbow, 28mm | nr | 0.24 | 3.48 | 5.78 | 1.39 | 10.65 |
| male elbow, 28mm × 1in | nr | 0.24 | 3.48 | 9.88 | 2.00 | 15.36 |
| female elbow, 28mm × 1in | nr | 0.24 | 3.48 | 10.01 | 2.02 | 15.51 |
| slow bend | nr | 0.24 | 3.48 | 10.01 | 2.02 | 15.51 |
| return bend | nr | 0.24 | 3.48 | 18.26 | 3.26 | 25.00 |
| obtuse elbow | nr | 0.24 | 3.48 | 3.73 | 1.08 | 8.29 |
| equal tee, 28mm | nr | 0.24 | 3.48 | 4.91 | 1.26 | 9.65 |

| | Unit | Hours | Hours £ | Mat'ls £ | O & P £ | Total £ |
|---|---|---|---|---|---|---|
| **Extra over 28mm diameter pipework (cont'd)** | | | | | | |
| tee, reduced branch largest end 28mm | nr | 0.28 | 4.06 | 7.26 | 1.70 | 13.02 |
| tee, one end and one branch reduced, largest end 28mm | nr | 0.28 | 4.06 | 9.07 | 1.97 | 15.10 |
| tee, both ends reduced, largest end 28mm | nr | 0.28 | 4.06 | 18.72 | 3.42 | 26.20 |
| sweep tee, 90 degrees, 28mm | nr | 0.28 | 4.06 | 17.76 | 3.27 | 25.09 |
| double sweep tee, 28mm | nr | 0.28 | 4.06 | 19.18 | 3.49 | 26.73 |
| cross equal tee, 28mm | nr | 0.28 | 4.06 | 19.66 | 3.56 | 27.28 |
| stop end, 28mm | nr | 0.24 | 3.48 | 3.98 | 1.12 | 8.58 |
| bent male union connector, 28mm × 1in | nr | 0.24 | 3.48 | 17.58 | 3.16 | 24.22 |
| bent female union connector, 28mm × 1in | nr | 0.24 | 3.48 | 17.58 | 3.16 | 24.22 |
| made bend | nr | 0.24 | 3.48 | 0.00 | 0.52 | 4.00 |
| 35mm diameter, to timber | m | 0.28 | 4.06 | 4.22 | 1.24 | 9.52 |
| 35mm diameter, plugged and screwed | m | 0.30 | 4.35 | 4.48 | 1.32 | 10.15 |
| extra over for | | | | | | |
| straight coupling | nr | 0.26 | 3.77 | 3.59 | 1.10 | 8.46 |
| reduced coupling, 35 × 22mm | nr | 0.26 | 3.77 | 6.98 | 1.61 | 12.36 |
| reduced coupling, 35 × 28mm | nr | 0.26 | 3.77 | 7.12 | 1.63 | 12.52 |
| adaptor coupling, 35mm × 1 1/4in | nr | 0.26 | 3.77 | 7.49 | 1.69 | 12.95 |
| flanged connector | nr | 0.26 | 3.77 | 54.60 | 8.76 | 67.13 |
| straight female connector, 35mm × 1 1/4in | nr | 0.26 | 3.77 | 12.55 | 2.45 | 18.77 |
| straight male connector, 35mm × 1 1/4in | nr | 0.26 | 3.77 | 11.72 | 2.32 | 17.81 |
| tank connector, 35mm × 1 1/4in | nr | 0.26 | 3.77 | 14.84 | 2.79 | 21.40 |
| reducer, 35 × 15mm | nr | 0.26 | 3.77 | 7.28 | 1.66 | 12.71 |
| reducer, 35 × 22mm | nr | 0.26 | 3.77 | 6.92 | 1.60 | 12.29 |
| reducer, 35 × 28mm | nr | 0.26 | 3.77 | 5.13 | 1.34 | 10.24 |

| | Unit | Hours | Hours £ | Mat'ls £ | O & P £ | Total £ |
|---|---|---|---|---|---|---|
| **Extra over 28mm diameter pipework (cont'd)** | | | | | | |
| female adaptor, 35mm × 1 1/4in | nr | 0.26 | 3.77 | 18.18 | 3.29 | 25.24 |
| male adaptor, 35mm × 1 1/4in | nr | 0.26 | 3.77 | 14.85 | 2.79 | 21.41 |
| adaptor, 35mm × 1 1/4in | nr | 0.26 | 3.77 | 4.16 | 1.19 | 9.12 |
| elbow, 35mm | nr | 0.26 | 3.77 | 7.70 | 1.72 | 13.19 |
| street elbow, 35mm | nr | 0.26 | 3.77 | 12.95 | 2.51 | 19.23 |
| male elbow, 35mm × 1 1/4in | nr | 0.26 | 3.77 | 13.97 | 2.66 | 20.40 |
| female elbow, 35mm × 1 1/4in | nr | 0.26 | 3.77 | 15.49 | 2.89 | 22.15 |
| equal tee, 35mm | nr | 0.30 | 4.35 | 12.53 | 2.53 | 19.41 |
| tee, reduced branch largest end 35mm | nr | 0.30 | 4.35 | 16.47 | 3.12 | 23.94 |
| tee, one end and one branch reduced, largest end 35mm | nr | 0.30 | 4.35 | 20.74 | 3.76 | 28.85 |
| tee, both ends reduced, largest end 35mm | nr | 0.30 | 4.35 | 26.19 | 4.58 | 35.12 |
| sweep tee, 90 degrees, 35mm | nr | 0.30 | 4.35 | 25.18 | 4.43 | 33.96 |
| stop end, 35mm | nr | 0.26 | 3.77 | 8.79 | 1.88 | 14.44 |
| bent male union adaptor, 35mm × 1 1/4in | nr | 0.26 | 3.77 | 28.67 | 4.87 | 37.31 |
| bent female union adaptor, 35mm × 1 1/4in | nr | 0.26 | 3.77 | 28.67 | 4.87 | 37.31 |
| made bend | nr | 0.26 | 3.77 | 0.00 | 0.57 | 4.34 |
| 42mm diameter, to timber | m | 0.30 | 4.35 | 9.28 | 2.04 | 15.67 |
| 42mm diameter, plugged and screwed | m | 0.32 | 4.64 | 9.48 | 2.12 | 16.24 |
| extra over for | | | | | | |
| straight coupling | nr | 0.28 | 4.06 | 5.99 | 1.51 | 11.56 |
| reduced coupling, 42 × 28mm | nr | 0.28 | 4.06 | 11.91 | 2.40 | 18.37 |
| reduced coupling, 42 × 35mm | nr | 0.28 | 4.06 | 10.46 | 2.18 | 16.70 |
| adaptor coupling, 42mm × 1 1/2in | nr | 0.28 | 4.06 | 9.51 | 2.04 | 15.61 |

**Extra over 42mm diameter pipework (cont'd)**

| | Unit | Hours | Hours £ | Mat'ls £ | O & P £ | Total £ |
|---|---|---|---|---|---|---|
| flanged connector | nr | 0.28 | 4.06 | 65.26 | 10.40 | 79.72 |
| straight female connector, 42mm × 1 1/2in | nr | 0.28 | 4.06 | 16.68 | 3.11 | 23.85 |
| straight male connector, 42mm × 1 1/2in | nr | 0.28 | 4.06 | 15.08 | 2.87 | 22.01 |
| tank connector, 42mm × 1 1/2in | nr | 0.28 | 4.06 | 19.46 | 3.53 | 27.05 |
| reducer, 42 × 15mm | nr | 0.28 | 4.06 | 11.18 | 2.29 | 17.53 |
| reducer, 42 × 22mm | nr | 0.28 | 4.06 | 12.48 | 2.48 | 19.02 |
| reducer, 42 × 28mm | nr | 0.28 | 4.06 | 11.76 | 2.37 | 18.19 |
| reducer, 42 × 35mm | nr | 0.28 | 4.06 | 9.66 | 2.06 | 15.78 |
| female adaptor, 42mm × 1 1/2in | nr | 0.28 | 4.06 | 22.90 | 4.04 | 31.00 |
| male adaptor, 42mm × 1 1/2in | nr | 0.28 | 4.06 | 20.52 | 3.69 | 28.27 |
| adaptor, 42mm × 1 1/2in | nr | 0.28 | 4.06 | 5.48 | 1.43 | 10.97 |
| elbow, 42mm | nr | 0.28 | 4.06 | 12.72 | 2.52 | 19.30 |
| street elbow, 42mm | nr | 0.28 | 4.06 | 18.24 | 3.35 | 25.65 |
| male elbow, 42mm × 1 1/2in | nr | 0.28 | 4.06 | 16.61 | 3.10 | 23.77 |
| equal tee, 42mm | nr | 0.32 | 4.64 | 20.10 | 3.71 | 28.45 |
| tee, reduced branch largest end 42mm | nr | 0.32 | 4.64 | 37.92 | 6.38 | 48.94 |
| tee, one end and one branch reduced, largest end 42mm | nr | 0.32 | 4.64 | 35.47 | 6.02 | 46.13 |
| tee, both ends reduced, largest end 42mm | nr | 0.32 | 4.64 | 26.19 | 4.62 | 35.45 |
| sweep tee, 90 degrees, 42mm | nr | 0.32 | 4.64 | 37.33 | 6.30 | 48.27 |
| stop end, 42mm | nr | 0.28 | 4.06 | 15.13 | 2.88 | 22.07 |
| bent male union adaptor, 42mm × 1 1/2in | nr | 0.28 | 4.06 | 46.64 | 7.61 | 58.31 |
| bent female union adaptor, 42mm × 1 1/2in | nr | 0.28 | 4.06 | 46.64 | 7.61 | 58.31 |
| made bend | nr | 0.28 | 4.06 | 0.00 | 0.61 | 4.67 |

| | Unit | Hours | Hours £ | Mat'ls £ | O & P £ | Total £ |
|---|---|---|---|---|---|---|
| 54mm diameter, to timber | m | 0.32 | 4.64 | 13.10 | 2.66 | 20.40 |
| 54mm diameter, plugged and screwed | m | 0.34 | 4.93 | 13.51 | 2.77 | 21.21 |
| extra over for | | | | | | |
| straight coupling | nr | 0.30 | 4.35 | 11.05 | 2.31 | 17.71 |
| reduced coupling, 54 × 35mm | nr | 0.30 | 4.35 | 18.33 | 3.40 | 26.08 |
| reduced coupling, 54 × 42mm | nr | 0.30 | 4.35 | 19.99 | 3.65 | 27.99 |
| straight female connector, 54mm × 2in | nr | 0.30 | 4.35 | 26.45 | 4.62 | 35.42 |
| straight male connector, 54mm × 2in | nr | 0.30 | 4.35 | 22.91 | 4.09 | 31.35 |
| tank connector, 54mm × 2in | nr | 0.30 | 4.35 | 29.73 | 5.11 | 39.19 |
| reducer, 54 × 15mm | nr | 0.30 | 4.35 | 24.68 | 4.35 | 33.38 |
| reducer, 54 × 22mm | nr | 0.30 | 4.35 | 24.18 | 4.28 | 32.81 |
| reducer, 54 × 28mm | nr | 0.30 | 4.35 | 21.26 | 3.84 | 29.45 |
| reducer, 54 × 35mm | nr | 0.30 | 4.35 | 20.27 | 3.69 | 28.31 |
| reducer, 54 × 42mm | nr | 0.30 | 4.35 | 17.49 | 3.28 | 25.12 |
| female adaptor, 54mm × 2in | nr | 0.30 | 4.35 | 27.57 | 4.79 | 36.71 |
| male adaptor, 54mm × 2in | nr | 0.30 | 4.35 | 27.57 | 4.79 | 36.71 |
| adaptor, 54mm × 2in | nr | 0.30 | 4.35 | 12.00 | 2.45 | 18.80 |
| elbow, 54mm | nr | 0.30 | 4.35 | 12.94 | 2.59 | 19.88 |
| male elbow, 54mm × 2in | nr | 0.30 | 4.35 | 26.25 | 4.59 | 35.19 |
| equal tee, 54mm | nr | 0.34 | 4.93 | 29.38 | 5.15 | 39.46 |
| tee, reduced branch largest end 54mm | nr | 0.34 | 4.93 | 40.51 | 6.82 | 52.26 |
| tee, one end and one branch reduced, largest end 54mm | nr | 0.34 | 3.72 | 60.87 | 9.69 | 74.28 |
| sweep tee, 90 degrees, 54mm | nr | 0.34 | 3.72 | 53.78 | 8.62 | 66.12 |
| stop end, 54mm | nr | 0.34 | 3.72 | 41.36 | 6.76 | 51.84 |
| bent union adaptor, 54mm × 54mm × 2in | nr | 0.34 | 4.93 | 21.12 | 3.91 | 29.96 |
| bent male union adaptor, 54mm × 2in | nr | 0.34 | 4.93 | 73.66 | 11.79 | 90.38 |
| bent female union adaptor, 54mm × 2in | nr | 0.34 | 4.93 | 73.66 | 11.79 | 90.38 |

| | Unit | Hours | Hours £ | Mat'ls £ | O & P £ | Total £ |
|---|---|---|---|---|---|---|
| **Stop valves** | | | | | | |
| Gunmetal stop valve with brass headwork, copper × copper | | | | | | |
| 15mm | nr | 0.20 | 2.90 | 5.03 | 1.19 | 9.12 |
| 22mm | nr | 0.22 | 3.19 | 9.39 | 1.89 | 14.47 |
| 28mm | nr | 0.28 | 4.06 | 26.67 | 4.61 | 35.34 |
| Dezincification-resistant stop valve, copper × copper | | | | | | |
| 15mm | nr | 0.20 | 2.90 | 12.62 | 2.33 | 17.85 |
| 22mm | nr | 0.22 | 3.19 | 21.88 | 3.76 | 28.83 |
| 28mm | nr | 0.28 | 4.06 | 36.45 | 6.08 | 46.59 |
| Gunmetal stop valve, copper × copper | | | | | | |
| 35mm | nr | 0.32 | 4.64 | 57.17 | 9.27 | 71.08 |
| 42mm | nr | 0.35 | 5.08 | 75.92 | 12.15 | 93.14 |
| 54mm | nr | 0.42 | 6.09 | 113.41 | 17.93 | 137.43 |
| Gunmetal lockshield stop valve with brass headwork, copper × copper | | | | | | |
| 22mm | nr | 0.22 | 3.19 | 14.50 | 2.65 | 20.34 |
| 28mm | nr | 0.28 | 4.06 | 26.59 | 4.60 | 35.25 |
| Gunmetal union stop valve with brass headwork, copper × copper union | | | | | | |
| 15mm | nr | 0.20 | 2.90 | 14.19 | 2.56 | 19.65 |
| 22mm | nr | 0.22 | 3.19 | 20.84 | 3.60 | 27.63 |
| Gunmetal double union lockshield stop valve with brass headwork, copper × copper union | | | | | | |
| 15mm | nr | 0.20 | 2.90 | 16.68 | 2.94 | 22.52 |
| 22mm | nr | 0.22 | 3.19 | 22.91 | 3.92 | 30.02 |
| 28mm | nr | 0.28 | 4.06 | 40.75 | 6.72 | 51.53 |

| | Unit | Hours | Hours £ | Mat'ls £ | O & P £ | Total £ |
|---|---|---|---|---|---|---|
| **Stop valves (cont'd)** | | | | | | |
| Dezincification-resistant double union lockshield stop valve, copper × copper | | | | | | |
| 15mm | nr | 0.20 | 2.90 | 26.85 | 4.46 | 34.21 |
| 22mm | nr | 0.22 | 3.19 | 32.68 | 5.38 | 41.25 |
| 28mm | nr | 0.28 | 4.06 | 60.43 | 9.67 | 74.16 |
| Dezincification-resistant double union stop valve with easy-clean cover, copper × copper | | | | | | |
| 15mm | nr | 0.20 | 2.90 | 28.39 | 4.69 | 35.98 |
| 22mm | nr | 0.22 | 3.19 | 35.43 | 5.79 | 44.41 |
| 28mm | nr | 0.28 | 4.06 | 65.89 | 10.49 | 80.44 |
| Gunmetal double union stop valve with easy-clean cover, copper × copper | | | | | | |
| 35mm | nr | 0.35 | 5.08 | 100.66 | 15.86 | 121.60 |
| 42mm | nr | 0.42 | 6.09 | 138.16 | 21.64 | 165.89 |
| 54mm | nr | 0.46 | 6.67 | 217.34 | 33.60 | 257.61 |
| Combined gunmetal stop valve and draincock with brass headwork, copper × copper | | | | | | |
| 15mm | nr | 0.20 | 2.90 | 27.99 | 4.63 | 35.52 |
| 22mm | nr | 0.22 | 3.19 | 34.39 | 5.64 | 43.22 |
| **Gate valves** | | | | | | |
| Gunmetal fullway gate valve, copper × copper | | | | | | |
| 15mm | nr | 0.20 | 2.90 | 15.74 | 2.80 | 21.44 |
| 22mm | nr | 0.22 | 3.19 | 18.24 | 3.21 | 24.64 |
| 28mm | nr | 0.28 | 4.06 | 25.39 | 4.42 | 33.87 |
| 35mm | nr | 0.32 | 4.64 | 56.65 | 9.19 | 70.48 |
| 42mm | nr | 0.35 | 5.08 | 70.83 | 11.39 | 87.29 |
| 54mm | nr | 0.42 | 6.09 | 102.74 | 16.32 | 125.15 |

| | Unit | Hours | Hours £ | Mat'ls £ | O & P £ | Total £ |
|---|---|---|---|---|---|---|
| **Plug cocks** | | | | | | |
| Nited | | | | | | |
| 15mm | nr | 0.20 | 2.90 | 11.68 | 2.19 | 16.77 |
| 22mm | nr | 0.20 | 2.90 | 17.58 | 3.07 | 23.55 |
| Nited with drop head | | | | | | |
| 15mm | nr | 0.25 | 3.63 | 13.27 | 2.53 | 19.43 |
| Nited with union outlet | | | | | | |
| 15mm | nr | 0.25 | 3.63 | 13.55 | 2.58 | 19.75 |
| 22mm | nr | 0.25 | 3.63 | 20.57 | 3.63 | 27.82 |
| Nited union stopcock with drop head | | | | | | |
| 15mm | nr | 0.25 | 3.63 | 15.36 | 2.85 | 21.83 |
| 22mm | nr | 0.25 | 3.63 | 22.14 | 3.86 | 29.63 |
| **Valves** | | | | | | |
| Radiator valve, plain finish, compression inlet × taper male union outlet | | | | | | |
| 8mm × 1/2in | nr | 0.30 | 4.35 | 16.94 | 3.19 | 24.48 |
| 10mm × 1/2in | nr | 0.30 | 4.35 | 18.44 | 3.42 | 26.21 |
| Radiator valve, chromium plated, compression inlet × taper male union outlet | | | | | | |
| 8mm × 1/2in | nr | 0.30 | 4.35 | 20.58 | 3.74 | 28.67 |
| 10mm × 1/2in | nr | 0.30 | 4.35 | 22.62 | 4.05 | 31.02 |
| Radiator valve, plain finish, compression inlet × taper male union outlet | | | | | | |
| 8mm × 1/2in | nr | 0.30 | 4.35 | 6.60 | 1.64 | 12.59 |
| 10mm × 1/2in | nr | 0.30 | 4.35 | 6.60 | 1.64 | 12.59 |

| | Unit | Hours | Hours £ | Mat'ls £ | O & P £ | Total £ |
|---|---|---|---|---|---|---|
| Radiator valve, nickel plated, compression inlet × taper male union outlet | | | | | | |
| 8mm × 1/2in | nr | 0.30 | 4.35 | 6.82 | 1.68 | 12.85 |
| 10mm × 1/2in | nr | 0.30 | 4.35 | 6.82 | 1.68 | 12.85 |
| **Copper pipe, dezincification-resistant compression joints and fittings, clips at 1250mm maximum centres** | | | | | | |
| 15mm diameter, to timber | m | 0.26 | 3.77 | 1.72 | 0.82 | 6.31 |
| 15mm diameter, plugged and screwed | m | 0.28 | 4.06 | 1.96 | 0.90 | 6.92 |
| extra over for | | | | | | |
| male coupling, 15mm × 3/4in | nr | 0.18 | 1.97 | 4.88 | 1.03 | 7.88 |
| tank coupling with backnut, 15mm × 1/2in | nr | 0.18 | 2.61 | 5.70 | 1.25 | 9.56 |
| female coupling, 15mm × 3/4in | nr | 0.18 | 2.61 | 4.39 | 1.05 | 8.05 |
| male elbow, 15mm × 1/2in | nr | 0.18 | 2.61 | 3.70 | 0.95 | 7.26 |
| male elbow, BSP parallel thread, 15mm × 1/2in | nr | 0.18 | 2.61 | 3.70 | 0.95 | 7.26 |
| female elbow, 15mm × 1/2in | nr | 0.18 | 2.61 | 3.95 | 0.98 | 7.54 |
| female elbow, 15mm × 3/4in | nr | 0.18 | 2.61 | 7.10 | 1.46 | 11.17 |
| female wall elbow, 15mm × 1/2in | nr | 0.18 | 2.61 | 5.76 | 1.26 | 9.63 |
| tee with reduced branch, 15 × 15 × 12mm | nr | 0.22 | 3.19 | 7.34 | 1.58 | 12.11 |
| tee, end reduced 15 × 12 × 12mm | nr | 0.22 | 3.19 | 5.24 | 1.26 | 9.69 |
| tee, end and branch reduced, 15 × 12 × 12mm | nr | 0.22 | 3.19 | 6.07 | 1.39 | 10.65 |
| tee, both ends reduced, 15 × 15 × 18mm | nr | 0.22 | 3.19 | 10.44 | 2.04 | 15.67 |
| tee, both ends reduced, 15 × 15 × 22mm | nr | 0.22 | 3.19 | 8.20 | 1.71 | 13.10 |
| female tee, 15mm × 1/2in × 15mm | nr | 0.22 | 3.19 | 7.48 | 1.60 | 12.27 |

| | Unit | Hours | Hours £ | Mat'ls £ | O & P £ | Total £ |
|---|---|---|---|---|---|---|
| **Extra over 15mm diameter pipework (cont'd)** | | | | | | |
| male tee, 15mm × 1/2in × 15mm | nr | 0.22 | 3.19 | 8.57 | 1.76 | 13.52 |
| cross tee, 15mm | nr | 0.22 | 3.19 | 10.07 | 1.99 | 15.25 |
| straight swivel connector, 15mm × 3/4in | nr | 0.18 | 2.61 | 6.49 | 1.37 | 10.47 |
| bent swivel connector, 15mm × 1/2in | nr | 0.18 | 2.61 | 4.82 | 1.11 | 8.54 |
| 18mm diameter, to timber | m | 0.21 | 3.05 | 2.45 | 0.82 | 6.32 |
| 18mm diameter, plugged and screwed | m | 0.23 | 3.34 | 2.71 | 0.91 | 6.95 |
| extra over for | | | | | | |
| straight coupling, 18mm | nr | 0.18 | 2.61 | 4.97 | 1.14 | 8.72 |
| male coupling, 18mm × 1/2in | nr | 0.18 | 2.61 | 4.02 | 0.99 | 7.62 |
| female coupling, 18mm × 1/2in | nr | 0.18 | 2.61 | 4.39 | 1.05 | 8.05 |
| elbow, 18mm | nr | 0.18 | 2.61 | 5.45 | 1.21 | 9.27 |
| male elbow, 18mm × 1/2in | nr | 0.18 | 2.61 | 5.45 | 1.21 | 9.27 |
| female elbow, 18mm × 1/2in | nr | 0.18 | 2.61 | 6.23 | 1.33 | 10.17 |
| tee, 18mm | nr | 0.22 | 3.19 | 7.83 | 1.65 | 12.67 |
| tee with reduced branch, 18 × 15 × 12mm | nr | 0.22 | 3.19 | 7.34 | 1.58 | 12.11 |
| tee with reduced branch, 18 × 18 × 15mm | nr | 0.22 | 3.19 | 9.41 | 1.89 | 14.49 |
| tee, end reduced 18 × 12 × 18mm | nr | 0.22 | 3.19 | 10.28 | 2.02 | 15.49 |
| tee, end reduced 18 × 15 × 18mm | nr | 0.22 | 3.19 | 9.41 | 1.89 | 14.49 |
| tee, end and branch reduced, 18 × 12 × 12mm | nr | 0.22 | 3.19 | 9.41 | 1.89 | 14.49 |
| tee, end and branch reduced, 18 × 15 × 12mm | nr | 0.22 | 3.19 | 9.41 | 1.89 | 14.49 |
| tee, end and branch reduced, 18 × 15 × 12mm | nr | 0.22 | 3.19 | 9.23 | 1.86 | 14.28 |
| reducing set, 18 × 12mm | nr | 0.18 | 2.61 | 2.43 | 0.76 | 5.80 |
| reducing set, 18 × 15mm | nr | 0.18 | 2.61 | 2.43 | 0.76 | 5.80 |

| | Unit | Hours | Hours £ | Mat'ls £ | O & P £ | Total £ |
|---|---|---|---|---|---|---|
| 22mm diameter, to timber | m | 0.22 | 3.19 | 2.81 | 0.90 | 6.90 |
| 22mm diameter, plugged and screwed | m | 0.24 | 3.48 | 3.12 | 0.99 | 7.59 |
| extra over for | | | | | | |
| male coupling, 22mm × 1/2in | nr | 0.18 | 2.61 | 4.02 | 0.99 | 7.62 |
| tank coupling with backnut, 22mm × 3/4in | nr | 0.18 | 2.61 | 6.09 | 1.31 | 10.01 |
| tank coupling with backnut, 22mm × 1in | nr | 0.18 | 2.61 | 10.14 | 1.91 | 14.66 |
| female coupling, 22mm × 1/2in | nr | 0.18 | 2.61 | 4.39 | 1.05 | 8.05 |
| female coupling, 22mm × 1in | nr | 0.18 | 2.61 | 6.36 | 1.35 | 10.32 |
| male elbow, 22mm × 3/4in | nr | 0.18 | 2.61 | 4.13 | 1.01 | 7.75 |
| male elbow, 22mm × 1in | nr | 0.18 | 2.61 | 8.44 | 1.66 | 12.71 |
| male elbow, BSP parallel thread, 22mm × 3/4in | nr | 0.18 | 2.61 | 4.13 | 1.01 | 7.75 |
| male elbow, BSP parallel thread, 22mm × 1in | nr | 0.18 | 2.61 | 8.44 | 1.66 | 12.71 |
| female elbow, 22mm × 3/4in | nr | 0.18 | 2.61 | 5.70 | 1.25 | 9.56 |
| female elbow, 22mm × 1in | nr | 0.18 | 2.61 | 9.14 | 1.76 | 13.51 |
| female wall elbow, 22mm × 3/4in | nr | 0.18 | 2.61 | 10.86 | 2.02 | 15.49 |
| tee with reduced branch, 22 × 22 × 12mm | nr | 0.22 | 3.19 | 10.93 | 2.12 | 16.24 |
| tee with reduced branch, 22 × 22 × 15mm | nr | 0.22 | 3.19 | 7.90 | 1.66 | 12.75 |
| tee, end reduced 22 × 15 × 22mm | nr | 0.22 | 3.19 | 9.18 | 1.86 | 14.23 |
| tee, end and branch reduced, 22 × 15 × 15mm | nr | 0.22 | 3.19 | 8.64 | 1.77 | 13.60 |
| tee, both ends reduced, 22 × 22 × 28mm | nr | 0.22 | 3.19 | 13.76 | 2.54 | 19.49 |
| female tee, 22mm × 1/2in × 22mm | nr | 0.22 | 3.19 | 9.42 | 1.89 | 14.50 |
| female tee, 22mm × 3/4in × 22mm | nr | 0.22 | 3.19 | 9.60 | 1.92 | 14.71 |
| male end tee, 22mm × 3/4in × 22mm | nr | 0.22 | 3.19 | 12.11 | 2.30 | 17.60 |

| | Unit | Hours | Hours £ | Mat'ls £ | O & P £ | Total £ |
|---|---|---|---|---|---|---|
| **Extra over 22mm diameter pipework (cont'd)** | | | | | | |
| straight swivel connector, 22mm × 3/4in | nr | 0.18 | 2.61 | 7.10 | 1.46 | 11.17 |
| bent swivel connector, 22mm × 3/4in | nr | 0.18 | 2.61 | 8.56 | 1.68 | 12.85 |
| male adaptor, 22mm × 1/2in | nr | 0.18 | 2.61 | 5.52 | 1.22 | 9.35 |
| male adaptor, 22mm × 3/4in | nr | 0.18 | 2.61 | 5.52 | 1.22 | 9.35 |
| female adaptor, 22mm × 1/2in | nr | 0.18 | 2.61 | 5.52 | 1.22 | 9.35 |
| female adaptor, 22mm × 3/4in | nr | 0.18 | 2.61 | 4.00 | 0.99 | 7.60 |
| reducing set, 22 × 12mm | nr | 0.18 | 2.61 | 2.98 | 0.84 | 6.43 |
| reducing set, 22 × 15mm | nr | 0.18 | 2.61 | 2.23 | 0.73 | 5.57 |
| reducing set, 22 × 18mm | nr | 0.18 | 2.61 | 2.56 | 0.78 | 5.95 |
| 28mm diameter, to timber | m | 0.23 | 3.34 | 3.46 | 1.02 | 7.81 |
| 28mm diameter, plugged and screwed | m | 0.25 | 3.63 | 3.70 | 1.10 | 8.42 |
| extra over for | | | | | | |
| straight coupling, 28mm | nr | 0.20 | 2.90 | 6.54 | 1.42 | 10.86 |
| male coupling, 28mm × 3/4in | nr | 0.20 | 2.90 | 6.81 | 1.46 | 11.17 |
| tank coupling with backnut, 28mm × 1in | nr | 0.20 | 2.90 | 10.46 | 2.00 | 15.36 |
| female coupling, 28mm × 1in | nr | 0.20 | 2.90 | 6.01 | 1.34 | 10.25 |
| elbow, 28mm | nr | 0.20 | 2.90 | 8.46 | 1.70 | 13.06 |
| male elbow, 28mm × 1in | nr | 0.20 | 2.90 | 8.44 | 1.70 | 13.04 |
| male elbow, BSP parallel thread, 28mm × 1in | nr | 0.20 | 2.90 | 7.74 | 1.60 | 12.24 |
| female elbow, 28mm × 1in | nr | 0.20 | 2.90 | 9.44 | 1.85 | 14.19 |
| tee, 28mm | nr | 0.24 | 3.48 | 13.48 | 2.54 | 19.50 |
| tee with reduced branch, 28 × 28 × 15mm | nr | 0.24 | 3.48 | 13.02 | 2.48 | 18.98 |
| tee with reduced branch, 28 × 28 × 22mm | nr | 0.24 | 3.48 | 13.02 | 2.48 | 18.98 |
| tee, end reduced 28 × 28 × 15mm | nr | 0.24 | 3.48 | 15.06 | 2.78 | 21.32 |
| tee, end reduced 28 × 28 × 22mm | nr | 0.24 | 3.48 | 14.27 | 2.66 | 20.41 |

| | Unit | Hours | Hours £ | Mat'ls £ | O & P £ | Total £ |
|---|---|---|---|---|---|---|
| **Extra over 28mm diameter pipework (cont'd)** | | | | | | |
| tee, both ends reduced, 28 × 28 × 22mm | nr | 0.24 | 3.48 | 14.12 | 2.64 | 20.24 |
| female tee, 28mm × 1/2in × 28mm | nr | 0.24 | 3.48 | 15.70 | 2.88 | 22.06 |
| female tee, 28mm × 1in × 28mm | nr | 0.24 | 3.48 | 15.70 | 2.88 | 22.06 |
| female adaptor, 28mm × 1in | nr | 0.20 | 2.90 | 9.08 | 1.80 | 13.78 |
| reducing set, 28 × 15mm | nr | 0.20 | 2.19 | 3.33 | 0.83 | 6.35 |
| reducing set, 28 × 22mm | nr | 0.20 | 2.19 | 3.33 | 0.83 | 6.35 |
| stop end, 28mm | nr | 0.20 | 2.19 | 6.84 | 1.35 | 10.38 |
| 35mm diameter, to timber | m | 0.28 | 4.06 | 7.27 | 1.70 | 13.03 |
| 35mm diameter, plugged and screwed | m | 0.30 | 4.35 | 7.51 | 1.78 | 13.64 |
| extra over for | | | | | | |
| straight coupling, 35mm | nr | 0.26 | 3.77 | 13.60 | 2.61 | 19.98 |
| male coupling, 35mm × 1 in | nr | 0.26 | 3.77 | 16.84 | 3.09 | 23.70 |
| male coupling, 35mm × 1 1/4in | nr | 0.26 | 3.77 | 10.33 | 2.12 | 16.22 |
| tank coupling with backnut, 35mm × 1 1/4in | nr | 0.26 | 3.77 | 12.44 | 2.43 | 18.64 |
| female coupling, 35mm × 1 1/4in | nr | 0.26 | 3.77 | 12.41 | 2.43 | 18.61 |
| elbow, 35mm | nr | 0.26 | 3.77 | 18.35 | 3.32 | 25.44 |
| male elbow, 35mm × 1 1/4in | nr | 0.26 | 3.77 | 17.07 | 3.13 | 23.97 |
| female elbow, 35mm × 1 1/4in | nr | 0.26 | 3.77 | 16.30 | 3.01 | 23.08 |
| tee, 35mm | nr | 0.30 | 4.35 | 23.89 | 4.24 | 32.48 |
| tee with reduced branch, 35 × 35 × 15mm | nr | 0.30 | 4.35 | 23.34 | 4.15 | 31.84 |
| tee with reduced branch, 35 × 35 × 22mm | nr | 0.30 | 4.35 | 23.34 | 4.15 | 31.84 |
| tee with reduced branch, 35 × 35 × 28mm | nr | 0.30 | 4.35 | 23.34 | 4.15 | 31.84 |
| reducing set, 35 × 15mm | nr | 0.26 | 3.77 | 5.71 | 1.42 | 10.90 |

| | Unit | Hours | Hours £ | Mat'ls £ | O & P £ | Total £ |
|---|---|---|---|---|---|---|
| **Extra over 35mm diameter pipework (cont'd)** | | | | | | |
| reducing set, 35 × 22mm | nr | 0.26 | 3.77 | 5.71 | 1.42 | 10.90 |
| reducing set, 35 × 28mm | nr | 0.26 | 3.77 | 5.71 | 1.42 | 10.90 |
| tank connector, 35mm | nr | 0.26 | 3.77 | 8.24 | 1.80 | 13.81 |
| stop end, 35mm | nr | 0.26 | 3.77 | 10.54 | 2.15 | 16.46 |
| 42mm diameter, to timber | m | 0.30 | 4.35 | 9.14 | 2.02 | 15.51 |
| 42mm diameter, plugged and screwed | m | 0.32 | 4.64 | 9.38 | 2.10 | 16.12 |
| extra over for | | | | | | |
| straight coupling, 42mm | nr | 0.28 | 4.06 | 17.88 | 3.29 | 25.23 |
| male coupling, 42mm × 1 1/2in | nr | 0.28 | 4.06 | 15.15 | 2.88 | 22.09 |
| tank coupling with backnut, 42mm × 1 1/2in | nr | 0.28 | 4.06 | 16.73 | 3.12 | 23.91 |
| female coupling, 42mm × 1 1/2in | nr | 0.28 | 4.06 | 16.70 | 3.11 | 23.87 |
| elbow, 42mm | nr | 0.28 | 4.06 | 24.86 | 4.34 | 33.26 |
| male elbow, 42mm × 1 1/2in | nr | 0.28 | 4.06 | 25.05 | 4.37 | 33.48 |
| female elbow, 42mm × 1 1/2in | nr | 0.28 | 4.06 | 25.06 | 4.37 | 33.49 |
| tee, 42mm | nr | 0.32 | 4.64 | 37.54 | 6.33 | 48.51 |
| tee with reduced branch, 42 × 42 × 22mm | nr | 0.32 | 4.64 | 36.07 | 6.11 | 46.82 |
| tee with reduced branch, 42 × 42 × 28mm | nr | 0.32 | 4.64 | 36.07 | 6.11 | 46.82 |
| reducing set, 42 × 15mm | nr | 0.28 | 4.06 | 10.41 | 2.17 | 16.64 |
| reducing set, 42 × 22mm | nr | 0.28 | 4.06 | 9.02 | 1.96 | 15.04 |
| reducing set, 42 × 28mm | nr | 0.28 | 4.06 | 9.02 | 1.96 | 15.04 |
| reducing set, 42mm × 35mm | nr | 0.28 | 4.06 | 9.02 | 1.96 | 15.04 |
| stop end, 42mm | nr | 0.28 | 4.06 | 17.54 | 3.24 | 24.84 |

| | Unit | Hours | Hours £ | Mat'ls £ | O & P £ | Total £ |
|---|---|---|---|---|---|---|
| 54mm diameter, to timber | m | 0.32 | 4.64 | 11.36 | 2.40 | 18.40 |
| 54mm diameter, plugged and screwed | m | 0.34 | 4.93 | 11.82 | 2.51 | 19.26 |
| extra over for | | | | | | |
| straight coupling, 54mm | nr | 0.30 | 4.35 | 26.74 | 4.66 | 35.75 |
| male coupling, 54mm × 2in | nr | 0.30 | 4.35 | 22.90 | 4.09 | 31.34 |
| tank coupling with backnut, 54mm × 2in | nr | 0.30 | 4.35 | 24.90 | 4.39 | 33.64 |
| female coupling, 54mm × 2in | nr | 0.30 | 4.35 | 24.49 | 4.33 | 33.17 |
| elbow, 54mm | nr | 0.30 | 4.35 | 42.79 | 7.07 | 54.21 |
| male elbow, 54mm × 2in | nr | 0.30 | 4.35 | 38.76 | 6.47 | 49.58 |
| tee, 54mm | nr | 0.34 | 4.93 | 60.31 | 9.79 | 75.03 |
| tee with reduced branch, 54 × 54 × 28mm | nr | 0.34 | 4.93 | 60.31 | 9.79 | 75.03 |
| reducing set, 54 × 22mm | nr | 0.30 | 4.35 | 17.28 | 3.24 | 24.87 |
| reducing set, 54 × 28mm | nr | 0.30 | 4.35 | 17.28 | 3.24 | 24.87 |
| reducing set, 54 × 35mm | nr | 0.30 | 4.35 | 15.05 | 2.91 | 22.31 |
| reducing set, 54 × 42mm | nr | 0.30 | 4.35 | 15.05 | 2.91 | 22.31 |

**Copper pipe, compression joints and fittings, clips at 1250mm centres**

| | Unit | Hours | Hours £ | Mat'ls £ | O & P £ | Total £ |
|---|---|---|---|---|---|---|
| 15mm diameter, to timber | m | 0.20 | 2.90 | 1.64 | 0.68 | 5.22 |
| 15mm diameter, plugged and screwed | m | 0.22 | 3.19 | 1.92 | 0.77 | 5.88 |
| extra over for | | | | | | |
| bent union radiator, 15mm × 1/2in | nr | 0.18 | 2.61 | 6.54 | 1.37 | 10.52 |
| air-release elbow, 15mm | nr | 0.18 | 2.61 | 8.44 | 1.66 | 12.71 |
| slow bend, 15mm | nr | 0.18 | 2.61 | 8.24 | 1.63 | 12.48 |
| male elbow, 15mm × 1/2in | nr | 0.18 | 2.61 | 2.29 | 0.74 | 5.64 |
| male elbow, 15mm × 3/4in | nr | 0.18 | 2.61 | 6.99 | 1.44 | 11.04 |
| male elbow, BSP parallel thread, 15mm × 1/2in | nr | 0.18 | 2.61 | 2.29 | 0.74 | 5.64 |
| female elbow, 15mm × 1/2in | nr | 0.18 | 2.61 | 3.52 | 0.92 | 7.05 |
| offset tee, 15 × 15 × 15mm | nr | 0.18 | 2.61 | 15.19 | 2.67 | 20.47 |

| | Unit | Hours | Hours £ | Mat'ls £ | O & P £ | Total £ |
|---|---|---|---|---|---|---|
| **Extra over 15mm diameter pipework (cont'd)** | | | | | | |
| male adaptor, 15mm × 1/2in | nr | 0.18 | 2.61 | 2.78 | 0.81 | 6.20 |
| reducing set, 15 × 8mm | nr | 0.18 | 2.61 | 1.89 | 0.68 | 5.18 |
| reducing set, 15 × 10mm | nr | 0.18 | 2.61 | 1.89 | 0.68 | 5.18 |
| reducing set, 15 × 12mm | nr | 0.18 | 2.61 | 1.89 | 0.68 | 5.18 |
| 22mm diameter, to timber | m | 0.21 | 3.05 | 2.46 | 0.83 | 6.33 |
| 22mm diameter, plugged and screwed | m | 0.23 | 3.34 | 2.64 | 0.90 | 6.87 |
| extra over for | | | | | | |
| lead coupling, 22mm | nr | 0.18 | 2.61 | 3.64 | 0.94 | 7.19 |
| air-release elbow, 22mm | nr | 0.18 | 2.61 | 11.17 | 2.07 | 15.85 |
| slow bend, 22mm | nr | 0.18 | 2.61 | 13.28 | 2.38 | 18.27 |
| male elbow, 22mm × 3/4in | nr | 0.18 | 2.61 | 2.96 | 0.84 | 6.41 |
| male elbow, 22mm × 1in | nr | 0.18 | 2.61 | 5.79 | 1.26 | 9.66 |
| male elbow, BSP parallel thread, 22mm × 3/4in | nr | 0.18 | 2.61 | 2.96 | 0.84 | 6.41 |
| male elbow, BSP parallel thread, 22mm × 1in | nr | 0.18 | 2.61 | 5.79 | 1.26 | 9.66 |
| female elbow, 22mm × 3/4in | nr | 0.18 | 2.61 | 5.09 | 1.16 | 8.86 |
| female elbow, 22mm × 1in | nr | 0.18 | 2.61 | 7.92 | 1.58 | 12.11 |
| bent cylinder connector, 22mm × 1in | nr | 0.18 | 2.61 | 11.64 | 2.14 | 16.39 |
| tee with reduced branch, 22 × 22 × 15mm | nr | 0.22 | 3.19 | 7.33 | 1.58 | 12.10 |
| tee, end reduced 22 × 15 × 22mm | nr | 0.22 | 3.19 | 8.53 | 1.76 | 13.48 |
| tee, end and branch reduced, 22 × 15 × 15mm | nr | 0.22 | 3.19 | 8.23 | 1.71 | 13.13 |
| straight swivel connector, 22mm × 1/2in | nr | 0.18 | 2.61 | 6.87 | 1.42 | 10.90 |
| bent swivel connector, 22mm × 1/2in | nr | 0.18 | 2.61 | 9.08 | 1.75 | 13.44 |
| offset tee, 22 × 22 × 15mm | nr | 0.22 | 3.19 | 22.35 | 3.83 | 29.37 |

| | Unit | Hours | Hours £ | Mat'ls £ | O & P £ | Total £ |
|---|---|---|---|---|---|---|
| 28mm diameter, to timber | m | 0.22 | 3.19 | 3.28 | 0.97 | 7.44 |
| 28mm diameter, plugged and screwed | m | 0.24 | 3.48 | 3.48 | 1.04 | 8.00 |
| extra over for | | | | | | |
| straight coupling, 28mm | nr | 0.20 | 2.90 | 5.63 | 1.28 | 9.81 |
| male coupling, 28mm × 1 1/4in | nr | 0.20 | 2.90 | 12.68 | 2.34 | 17.92 |
| female coupling, 28mm × 3/4in | nr | 0.20 | 2.90 | 7.39 | 1.54 | 11.83 |
| female coupling, 28mm × 1in | nr | 0.20 | 2.90 | 4.46 | 1.10 | 8.46 |
| slow bend, 28mm | nr | 0.20 | 2.90 | 17.88 | 3.12 | 23.90 |
| male elbow, 28mm × 1in | nr | 0.20 | 2.90 | 5.79 | 1.30 | 9.99 |
| male elbow, BSP parallel thread, 28mm × 1in | nr | 0.20 | 2.90 | 7.19 | 1.51 | 11.60 |
| female elbow, 28mm × 1in | nr | 0.20 | 2.90 | 8.95 | 1.78 | 13.63 |
| tee, 28 × 28 × 28mm | nr | 0.24 | 3.48 | 12.95 | 2.46 | 18.89 |
| **Stop valves** | | | | | | |
| Gunmetal stop valve with brass headwork, copper × copper | | | | | | |
| 15mm | nr | 0.19 | 2.76 | 6.90 | 1.45 | 11.10 |
| 22mm | nr | 0.21 | 3.05 | 12.13 | 2.28 | 17.45 |
| 28mm | nr | 0.27 | 3.92 | 31.63 | 5.33 | 40.88 |
| Gunmetal lockshield stop valve with brass headwork, copper × copper | | | | | | |
| 15mm | nr | 0.19 | 2.76 | 15.00 | 2.66 | 20.42 |
| 22mm | nr | 0.21 | 3.05 | 21.17 | 3.63 | 27.85 |
| Dezincification-resistant stop valve, copper × copper | | | | | | |
| 15mm | nr | 0.19 | 2.76 | 16.08 | 2.83 | 21.66 |
| 22mm | nr | 0.21 | 3.05 | 26.40 | 4.42 | 33.86 |
| 28mm | nr | 0.27 | 3.92 | 43.69 | 7.14 | 54.75 |

| | Unit | Hours | Hours £ | Mat'ls £ | O & P £ | Total £ |
|---|---|---|---|---|---|---|
| Dezincification-resistant lockshield stop valve, copper × copper | | | | | | |
| 15mm | nr | 0.19 | 2.76 | 18.49 | 3.19 | 24.43 |
| 22mm | nr | 0.21 | 3.05 | 28.95 | 4.80 | 36.79 |
| Gunmetal stop valve with brass headwork, polyethylene × copper | | | | | | |
| 20 × 15mm | nr | 0.19 | 2.76 | 16.38 | 2.87 | 22.01 |
| 25 × 15mm | nr | 0.21 | 3.05 | 27.94 | 4.65 | 35.63 |
| 25 × 22mm | nr | 0.27 | 3.92 | 31.51 | 5.31 | 40.74 |
| **Gate valves** | | | | | | |
| Brass gate valve, copper × copper | | | | | | |
| 15mm | nr | 0.19 | 2.76 | 11.26 | 2.10 | 16.12 |
| 22mm | nr | 0.21 | 3.05 | 13.13 | 2.43 | 18.60 |
| 28mm | nr | 0.27 | 3.92 | 22.92 | 4.03 | 30.86 |
| Brass lockshield gate valve, copper × copper | | | | | | |
| 15mm | nr | 0.19 | 2.76 | 11.36 | 2.12 | 16.23 |
| 22mm | nr | 0.21 | 3.05 | 13.39 | 2.47 | 18.90 |
| 28mm | nr | 0.27 | 3.92 | 24.27 | 4.23 | 32.41 |
| Gunmetal fullway gate valve, copper × copper | | | | | | |
| 15mm | nr | 0.19 | 2.76 | 19.57 | 3.35 | 25.67 |
| 22mm | nr | 0.21 | 3.05 | 23.06 | 3.92 | 30.02 |
| 28mm | nr | 0.27 | 3.92 | 31.34 | 5.29 | 40.54 |
| 35mm | nr | 0.36 | 5.22 | 45.46 | 7.60 | 58.28 |
| 42mm | nr | 0.50 | 7.25 | 62.98 | 10.53 | 80.76 |
| 54mm | nr | 0.60 | 8.70 | 86.69 | 14.31 | 109.70 |

| | Unit | Hours | Hours £ | Mat'ls £ | O & P £ | Total £ |
|---|---|---|---|---|---|---|
| Gunmetal lockshield gate valve, copper × copper | | | | | | |
| 15mm | nr | 0.19 | 2.76 | 20.11 | 3.43 | 26.29 |
| 22mm | nr | 0.21 | 3.05 | 23.85 | 4.03 | 30.93 |
| 28mm | nr | 0.27 | 3.92 | 35.59 | 5.93 | 45.43 |
| 35mm | nr | 0.36 | 5.22 | 47.97 | 7.98 | 61.17 |
| 42mm | nr | 0.50 | 7.25 | 66.06 | 11.00 | 84.31 |
| 54mm | nr | 0.60 | 8.70 | 91.12 | 14.97 | 114.79 |
| **Check valves** | | | | | | |
| Single check valve with female BSP parallel threads | | | | | | |
| 1/2in | nr | 0.19 | 2.76 | 14.07 | 2.52 | 19.35 |
| 3/4in | nr | 0.21 | 3.05 | 22.44 | 3.82 | 29.31 |
| 1in | nr | 0.27 | 3.92 | 30.63 | 5.18 | 39.73 |
| Double check valve with female BSP parallel threads | | | | | | |
| 1/2in | nr | 0.19 | 2.76 | 15.93 | 2.80 | 21.49 |
| 3/4in | nr | 0.21 | 3.05 | 32.80 | 5.38 | 41.22 |
| 1in | nr | 0.27 | 3.92 | 49.17 | 7.96 | 61.05 |
| **Drainage fittings** | | | | | | |
| Dezincification-resistant stop valve adaptor, copper × copper | | | | | | |
| 15mm | nr | 0.20 | 2.90 | 10.96 | 2.08 | 15.94 |
| Draw-off coupling, copper × copper | | | | | | |
| 15mm | nr | 0.20 | 2.90 | 10.01 | 1.94 | 14.85 |
| Dezincification-resistant draw-off coupling, copper × copper | | | | | | |
| 15mm | nr | 0.20 | 2.90 | 13.82 | 2.51 | 19.23 |
| 22mm | nr | 0.25 | 3.63 | 15.99 | 2.94 | 22.56 |

| | Unit | Hours | Hours £ | Mat'ls £ | O & P £ | Total £ |
|---|---|---|---|---|---|---|
| Dezincification-resistant draw-off elbow, copper × parallel main thread | | | | | | |
| 15mm × 1/2in | nr | 0.20 | 2.90 | 10.55 | 2.02 | 15.47 |
| 22mm × 3/4in | nr | 0.25 | 3.63 | 11.84 | 2.32 | 17.78 |
| 28mm × 1in | nr | 0.30 | 4.35 | 19.32 | 3.55 | 27.22 |
| Brass draw-off elbow, copper × copper | | | | | | |
| 15mm | nr | 0.20 | 2.90 | 10.24 | 1.97 | 15.11 |
| 22mm | nr | 0.25 | 3.63 | 11.74 | 2.30 | 17.67 |
| 28mm | nr | 0.30 | 4.35 | 18.09 | 3.37 | 25.81 |
| Dezincification-resistant draw-off elbow, copper × copper | | | | | | |
| 15mm | nr | 0.20 | 2.90 | 15.17 | 2.71 | 20.78 |
| 22mm | nr | 0.25 | 3.63 | 18.70 | 3.35 | 25.67 |
| **Servicing valves** | | | | | | |
| Straight pattern, copper × copper | | | | | | |
| 15 × 15mm | nr | 0.20 | 2.90 | 5.38 | 1.24 | 9.52 |
| 22 × 22mm | nr | 0.25 | 3.63 | 11.94 | 2.33 | 17.90 |
| Straight pattern, copper × swivel end | | | | | | |
| 15mm × 1/2in | nr | 0.20 | 2.90 | 6.08 | 1.35 | 10.33 |
| **Appliance valves** | | | | | | |
| Chromium-plated stop valve | | | | | | |
| 15mm × 3/4in | nr | 0.20 | 2.90 | 4.32 | 1.08 | 8.30 |
| Brass tee valve | | | | | | |
| 15mm × 3/4in | nr | 0.20 | 2.90 | 5.85 | 1.31 | 10.06 |

| | Unit | Hours | Hours £ | Mat'ls £ | O & P £ | Total £ |
|---|---|---|---|---|---|---|
| **Cold water storage tanks** | | | | | | |
| Galvanised steel cistern with cover, reference | | | | | | |
| SC10, 18 litres | nr | 0.60 | 8.70 | 49.22 | 8.69 | 66.61 |
| SC15, 36 litres | nr | 0.65 | 9.43 | 62.02 | 10.72 | 82.16 |
| SC20, 54 litres | nr | 0.65 | 9.43 | 64.77 | 11.13 | 85.32 |
| SC25, 68 litres | nr | 0.70 | 10.15 | 68.26 | 11.76 | 90.17 |
| SC30, 86 litres | nr | 0.75 | 10.88 | 76.07 | 13.04 | 99.99 |
| SC40, 114 litres | nr | 0.80 | 11.60 | 82.19 | 14.07 | 107.86 |
| SC50, 154 litres | nr | 0.85 | 12.33 | 121.91 | 20.14 | 154.37 |
| Cutting holes for connectors | | | | | | |
| 13mm | nr | 0.10 | 1.45 | 0.00 | 0.22 | 1.67 |
| 18mm | nr | 0.12 | 1.74 | 0.00 | 0.26 | 2.00 |
| 25mm | nr | 0.15 | 2.18 | 0.00 | 0.33 | 2.50 |
| 32mm | nr | 0.20 | 2.90 | 0.00 | 0.44 | 3.34 |
| Plastic cistern with lid, reference | | | | | | |
| PC4, 18 litres | nr | 0.60 | 8.70 | 6.82 | 2.33 | 17.85 |
| PC15, 68 litres | nr | 0.70 | 10.15 | 16.44 | 3.99 | 30.58 |
| PC20, 91 litres | nr | 0.75 | 10.88 | 16.75 | 4.14 | 31.77 |
| PC25, 114 litres | nr | 0.80 | 11.60 | 22.28 | 5.08 | 38.96 |
| PC40, 182 litres | nr | 0.90 | 13.05 | 39.83 | 7.93 | 60.81 |
| PC50, 227 litres | nr | 1.00 | 14.50 | 40.13 | 8.19 | 62.82 |
| Cutting holes for connectors | | | | | | |
| 13mm | nr | 0.10 | 1.45 | 0.00 | 0.22 | 1.67 |
| 18mm | nr | 0.12 | 1.74 | 0.00 | 0.26 | 2.00 |
| 25mm | nr | 0.15 | 2.18 | 0.00 | 0.33 | 2.50 |
| 32mm | nr | 0.20 | 2.90 | 0.00 | 0.44 | 3.34 |

| | Unit | Hours | Hours £ | Mat'ls £ | O & P £ | Total £ |
|---|---|---|---|---|---|---|
| **Hot water copper cylinders** | | | | | | |
| Direct cylinders, insulated, reference | | | | | | |
| ref 5, 98 litres | nr | 0.50 | 7.25 | 124.34 | 19.74 | 151.33 |
| ref 7, 120 litres | nr | 0.55 | 7.98 | 141.93 | 22.49 | 172.39 |
| ref 8, 148 litres | nr | 0.65 | 9.43 | 156.08 | 24.83 | 190.33 |
| ref 9, 166 litres | nr | 0.85 | 12.33 | 170.37 | 27.40 | 210.10 |
| Indirect cylinders, insulated, reference | | | | | | |
| ref 2, 96 litres | nr | 0.50 | 7.25 | 98.48 | 15.86 | 121.59 |
| ref 3, 114 litres | nr | 0.55 | 7.98 | 107.11 | 17.26 | 132.35 |
| ref 7, 117 litres | nr | 0.55 | 7.98 | 112.62 | 18.09 | 138.68 |
| ref 8, 140 litres | nr | 0.65 | 9.43 | 128.41 | 20.68 | 158.51 |
| ref 9, 162 litres | nr | 0.70 | 10.15 | 159.80 | 25.49 | 195.44 |
| **Insulation** | | | | | | |
| Preformed pipe lagging, fire-retardant foam 13mm thick for pipe size | | | | | | |
| 15mm | m | 0.08 | 1.16 | 1.11 | 0.34 | 2.61 |
| 22mm | m | 0.09 | 1.31 | 1.45 | 0.41 | 3.17 |
| 28mm | m | 0.10 | 1.45 | 1.72 | 0.48 | 3.65 |
| Preformed pipe lagging, fire-retardant foam 25mm thick for pipe size | | | | | | |
| 15mm | m | 0.10 | 1.45 | 4.27 | 0.86 | 6.58 |
| 22mm | m | 0.11 | 1.60 | 4.70 | 0.94 | 7.24 |
| 28mm | m | 0.12 | 1.74 | 5.40 | 1.07 | 8.21 |
| 50mm glass-fibre filled insulating jacket, fixing bands to tank size | | | | | | |
| 445 × 305 × 300mm | nr | 0.35 | 5.08 | 4.26 | 1.40 | 10.74 |
| 495 × 368 × 362mm | nr | 0.40 | 5.80 | 5.21 | 1.65 | 12.66 |
| 630 × 450 × 420mm | nr | 0.45 | 6.53 | 6.03 | 1.88 | 14.44 |
| 665 × 490 × 515mm | nr | 0.50 | 7.25 | 7.07 | 2.15 | 16.47 |

| | Unit | Hours | Hours £ | Mat'ls £ | O & P £ | Total £ |
|---|---|---|---|---|---|---|
| **T10 GAS/OIL-FIRED BOILERS** | | | | | | |
| **Gas-fired wall-mounted boilers for domestic central heating and indirect hot water** | | | | | | |
| Balanced flue, output | | | | | | |
| 30,000 BTU | nr | 4.80 | 69.60 | 417.57 | 73.08 | 560.25 |
| 40,000 BTU | nr | 4.80 | 69.60 | 481.22 | 82.62 | 633.44 |
| 50,000 BTU | nr | 4.80 | 69.60 | 548.23 | 92.67 | 710.50 |
| 60,000 BTU | nr | 4.80 | 69.60 | 635.42 | 105.75 | 810.77 |
| Fan flue, output | | | | | | |
| 30,000 BTU | nr | 4.20 | 60.90 | 445.56 | 75.97 | 582.43 |
| 40,000 BTU | nr | 4.20 | 60.90 | 501.79 | 84.40 | 647.09 |
| 50,000 BTU | nr | 4.20 | 60.90 | 547.51 | 91.26 | 699.67 |
| 60,000 BTU | nr | 4.20 | 60.90 | 592.59 | 98.02 | 751.51 |
| 70,000 BTU | nr | 4.20 | 60.90 | 689.09 | 112.50 | 862.49 |
| 80,000 BTU | nr | 4.20 | 60.90 | 1059.09 | 168.00 | 1287.99 |
| **Gas-fired floor-standing boilers for domestic central heating and indirect hot water** | | | | | | |
| Fanned balanced flue, output | | | | | | |
| 40,000 BTU | nr | 4.80 | 69.60 | 552.20 | 93.27 | 715.07 |
| 50,000 BTU | nr | 4.80 | 69.60 | 566.30 | 95.39 | 731.29 |
| 60,000 BTU | nr | 4.80 | 69.60 | 586.64 | 98.44 | 754.68 |
| 70,000 BTU | nr | 4.80 | 69.60 | 623.27 | 103.93 | 796.80 |
| 80,000 BTU | nr | 4.80 | 69.60 | 856.33 | 138.89 | 1064.82 |
| Fanned chimney flue, output | | | | | | |
| 40,000 BTU | nr | 4.20 | 60.90 | 435.68 | 74.49 | 571.07 |
| 50,000 BTU | nr | 4.20 | 60.90 | 476.51 | 80.61 | 618.02 |
| 60,000 BTU | nr | 4.20 | 60.90 | 524.96 | 87.88 | 673.74 |
| 80,000 BTU | nr | 4.20 | 60.90 | 684.56 | 111.82 | 857.28 |
| 80,000 BTU | nr | 4.20 | 60.90 | 872.97 | 140.08 | 1073.95 |

| | Unit | Hours | Hours £ | Mat'ls £ | O & P £ | Total £ |
|---|---|---|---|---|---|---|
| **Gas-fired wall-mounted combination boiler for domestic central heating and hot water** | | | | | | |
| Fan flue, output | | | | | | |
| 30,400-81,900 BTU | nr | 4.80 | 69.60 | 651.68 | 108.19 | 829.47 |
| 35,500-95,000 BTU | nr | 4.80 | 69.60 | 696.68 | 114.94 | 881.22 |
| Condensing fan flue | | | | | | |
| 25,200-92,800 BTU | nr | 4.80 | 69.60 | 815.68 | 132.79 | 1018.07 |
| 25,900-93,800 BTU | nr | 4.80 | 69.60 | 843.68 | 136.99 | 1050.27 |
| Conventional flue | | | | | | |
| 31,000-82,000 BTU | nr | 4.80 | 69.60 | 666.68 | 110.44 | 846.72 |
| **Gas-fired floor-standing combination boiler for domestic central heating and hot water** | | | | | | |
| Fan flue, output | | | | | | |
| 37,500-81,900 BTU | nr | 4.80 | 69.60 | 1074.25 | 171.58 | 1315.43 |
| **Oil-fired boiler for domestic central heating and hot water** | | | | | | |
| Balanced flue, output | | | | | | |
| 40,000-48,000 BTU | nr | 4.80 | 69.60 | 754.00 | 10.44 | 80.04 |
| 50,000-65,000 BTU | nr | 4.80 | 69.60 | 843.75 | 191.15 | 1465.50 |
| 75,000-85,000 BTU | nr | 4.80 | 69.60 | 942.25 | 206.83 | 1585.68 |
| 88,000-110,000 BTU | nr | 4.80 | 69.60 | 1056.63 | 227.84 | 1746.80 |
| Conventional flue, output | | | | | | |
| 52,000 BTU | nr | 4.80 | 69.60 | 754.00 | 123.54 | 947.14 |
| 70,000 BTU | nr | 4.80 | 69.60 | 843.75 | 137.00 | 1050.35 |
| 90,000 BTU | nr | 4.80 | 69.60 | 942.25 | 151.78 | 1163.63 |
| 120,000 BTU | nr | 4.80 | 69.60 | 1056.63 | 168.93 | 1295.16 |

| | Unit | Hours | Hours £ | Mat'ls £ | O & P £ | Total £ |
|---|---|---|---|---|---|---|
| **Oil-fired combination boiler for domestic central heating and hot water** | | | | | | |
| 50,000-65,000 BTU | nr | 4.80 | 69.60 | 1204.75 | 191.15 | 1465.50 |
| 75,000-85,000 BTU | nr | 4.80 | 69.60 | 1309.25 | 206.83 | 1585.68 |
| 88,000-110,000 BTU | nr | 4.80 | 69.60 | 1449.36 | 227.84 | 1746.80 |
| Conventional flue, output | | | | | | |
| 52,000 BTU | nr | 4.80 | 69.60 | 1138.51 | 181.22 | 1389.33 |
| 70,000 BTU | nr | 4.80 | 69.60 | 1204.75 | 191.15 | 1465.50 |
| 90,000 BTU | nr | 4.80 | 69.60 | 1309.25 | 206.83 | 1585.68 |
| 120,000 BTU | nr | 4.80 | 69.60 | 1449.36 | 227.84 | 1746.80 |
| **Oil storage tanks** | | | | | | |
| Standard domestic pattern plastic oil storage tank | | | | | | |
| 1380 × 1780 × 890mm capacity 1200 litres | nr | 1.50 | 21.75 | 170.48 | 28.83 | 221.06 |
| 745 × 1700 × 1380mm capacity 1225 litres | nr | 1.50 | 21.75 | 241.93 | 39.55 | 303.23 |
| 1190mm diameter × 1400 high, capacity 1300 litres | nr | 1.50 | 21.75 | 170.86 | 28.89 | 221.50 |
| 1130 × 2055 × 1145mm capacity 1800 litres | nr | 1.75 | 25.38 | 299.49 | 48.73 | 373.59 |
| 1265 × 2250 × 1320mm capacity 2500 litres | nr | 2.00 | 29.00 | 373.05 | 60.31 | 462.36 |
| 1640mm diameter × 1750 high, capacity 3600 litres | nr | 2.50 | 36.25 | 411.42 | 67.15 | 514.82 |
| 2480mm diameter × 2640 high, capacity 10,000 litres | nr | 3.00 | 43.50 | 1189.64 | 184.97 | 1418.11 |
| Standard domestic pattern plastic bunded oil storage tank | | | | | | |
| 1380 × 1780 × 890mm capacity 1200 litres | nr | 1.50 | 21.75 | 228.27 | 37.50 | 287.52 |
| 745 × 1700 × 1380mm capacity 1225 litres | nr | 1.50 | 21.75 | 299.34 | 48.16 | 369.25 |

| | Unit | Hours | Hours £ | Mat'ls £ | O & P £ | Total £ |
|---|---|---|---|---|---|---|
| 1190mm diameter × 1400 high, capacity 1300 litres | nr | 1.50 | 21.75 | 699.75 | 108.23 | 829.73 |
| 1130 × 2055 × 1145mm capacity 1800 litres | nr | 1.75 | 25.38 | 730.58 | 113.39 | 869.35 |
| 1265 × 2250 × 1320mm capacity 2500 litres | nr | 2.00 | 29.00 | 811.36 | 126.05 | 966.41 |

| | Unit | Hours | Hours £ | Mat'ls £ | O & P £ | Total £ |
|---|---|---|---|---|---|---|
| **T30 RADIATORS** | | | | | | |
| **Single convector radiator fixed to concealed brackets plugged and screwed to brickwork** | | | | | | |
| 300mm high, length | | | | | | |
| 400mm | nr | 0.90 | 13.05 | 14.75 | 4.17 | 31.97 |
| 800mm | nr | 1.05 | 15.23 | 24.49 | 5.96 | 45.67 |
| 1200mm | nr | 1.15 | 16.68 | 35.24 | 7.79 | 59.70 |
| 1600mm | nr | 1.30 | 18.85 | 45.98 | 9.72 | 74.55 |
| 2000mm | nr | 1.40 | 20.30 | 55.73 | 11.40 | 87.43 |
| 450mm high, length | | | | | | |
| 500mm | nr | 1.00 | 14.50 | 14.43 | 4.34 | 33.27 |
| 1000mm | nr | 1.15 | 16.68 | 25.19 | 6.28 | 48.14 |
| 1400mm | nr | 1.25 | 18.13 | 33.62 | 7.76 | 59.51 |
| 1800mm | nr | 1.40 | 20.30 | 48.53 | 10.32 | 79.15 |
| 2000mm | nr | 1.50 | 21.75 | 64.56 | 12.95 | 99.26 |
| 600mm high, length | | | | | | |
| 500mm | nr | 1.10 | 15.95 | 17.80 | 5.06 | 38.81 |
| 1000mm | nr | 1.25 | 18.13 | 30.60 | 7.31 | 56.03 |
| 1400mm | nr | 1.35 | 19.58 | 41.64 | 9.18 | 70.40 |
| 1800mm | nr | 1.50 | 21.75 | 60.58 | 12.35 | 94.68 |
| 2000mm | nr | 1.60 | 23.20 | 80.54 | 15.56 | 119.30 |
| 700mm high, length | | | | | | |
| 500mm | nr | 1.20 | 17.40 | 19.80 | 5.58 | 42.78 |
| 1000mm | nr | 1.35 | 19.58 | 34.60 | 8.13 | 62.30 |
| 1400mm | nr | 1.45 | 21.03 | 63.64 | 12.70 | 97.36 |
| 1800mm | nr | 1.60 | 23.20 | 82.58 | 15.87 | 121.65 |
| 2000mm | nr | 1.70 | 24.65 | 91.54 | 17.43 | 133.62 |

| | Unit | Hours | Hours £ | Mat'ls £ | O & P £ | Total £ |
|---|---|---|---|---|---|---|
| **Double convector radiator fixed to concealed brackets plugged and screwed to brickwork** | | | | | | |
| 300mm high, length | | | | | | |
| 400mm | nr | 0.90 | 13.05 | 24.75 | 5.67 | 43.47 |
| 800mm | nr | 1.05 | 15.23 | 43.49 | 8.81 | 67.52 |
| 1200mm | nr | 1.15 | 16.68 | 62.24 | 11.84 | 90.75 |
| 1600mm | nr | 1.30 | 18.85 | 83.98 | 15.42 | 118.25 |
| 2000mm | nr | 1.40 | 20.30 | 103.73 | 18.60 | 142.63 |
| 450mm high, length | | | | | | |
| 500mm | nr | 1.00 | 14.50 | 23.43 | 5.69 | 43.62 |
| 1000mm | nr | 1.15 | 16.68 | 43.19 | 8.98 | 68.84 |
| 1400mm | nr | 1.25 | 18.13 | 66.62 | 12.71 | 97.46 |
| 1800mm | nr | 1.40 | 20.30 | 107.53 | 19.17 | 147.00 |
| 2000mm | nr | 1.50 | 21.75 | 118.56 | 21.05 | 161.36 |
| 600mm high, length | | | | | | |
| 500mm | nr | 1.10 | 15.95 | 28.10 | 6.61 | 50.66 |
| 1000mm | nr | 1.25 | 18.13 | 53.19 | 10.70 | 82.01 |
| 1400mm | nr | 1.35 | 19.58 | 82.68 | 15.34 | 117.59 |
| 1800mm | nr | 1.50 | 21.75 | 133.28 | 23.25 | 178.28 |
| 2000mm | nr | 1.60 | 23.20 | 147.96 | 25.67 | 196.83 |
| 700mm high, length | | | | | | |
| 500mm | nr | 1.20 | 17.40 | 31.80 | 7.38 | 56.58 |
| 1000mm | nr | 1.35 | 19.58 | 60.60 | 12.03 | 92.20 |
| 1400mm | nr | 1.45 | 21.03 | 112.64 | 20.05 | 153.71 |
| 1800mm | nr | 1.60 | 23.20 | 150.58 | 26.07 | 199.85 |
| 2000mm | nr | 1.70 | 24.65 | 165.54 | 28.53 | 218.72 |

| | Unit | Hours | Hours £ | Mat'ls £ | O & P £ | Total £ |
|---|---|---|---|---|---|---|
| **Double panel radiator fixed to concealed brackets plugged and screwed to brickwork** | | | | | | |
| 300mm high, length | | | | | | |
| 400mm | nr | 0.90 | 13.05 | 19.75 | 4.92 | 37.72 |
| 800mm | nr | 1.05 | 15.23 | 34.49 | 7.46 | 57.17 |
| 1200mm | nr | 1.15 | 16.68 | 50.24 | 10.04 | 76.95 |
| 1600mm | nr | 1.30 | 18.85 | 65.98 | 12.72 | 97.55 |
| 2000mm | nr | 1.40 | 20.30 | 80.73 | 15.15 | 116.18 |
| 450mm high, length | | | | | | |
| 500mm | nr | 1.00 | 14.50 | 19.43 | 5.09 | 39.02 |
| 1000mm | nr | 1.15 | 16.68 | 34.19 | 7.63 | 58.49 |
| 1400mm | nr | 1.25 | 18.13 | 52.62 | 10.61 | 81.36 |
| 1800mm | nr | 1.40 | 20.30 | 84.53 | 15.72 | 120.55 |
| 2000mm | nr | 1.50 | 21.75 | 93.56 | 17.30 | 132.61 |
| 600mm high, length | | | | | | |
| 500mm | nr | 1.10 | 15.95 | 23.10 | 5.86 | 44.91 |
| 1000mm | nr | 1.25 | 18.13 | 42.19 | 9.05 | 69.36 |
| 1400mm | nr | 1.35 | 19.58 | 65.68 | 12.79 | 98.04 |
| 1800mm | nr | 1.50 | 21.75 | 105.28 | 19.05 | 146.08 |
| 2000mm | nr | 1.60 | 23.20 | 116.96 | 21.02 | 161.18 |
| 700mm high, length | | | | | | |
| 500mm | nr | 1.20 | 17.40 | 29.80 | 7.08 | 54.28 |
| 1000mm | nr | 1.35 | 19.58 | 57.60 | 11.58 | 88.75 |
| 1400mm | nr | 1.45 | 21.03 | 89.64 | 16.60 | 127.26 |
| 1800mm | nr | 1.60 | 23.20 | 119.58 | 21.42 | 164.20 |
| 2000mm | nr | 1.70 | 24.65 | 132.54 | 23.58 | 180.77 |

## ALTERATIONS AND REPAIRS

### Rainwater installation

| | Unit | Hours | Hours £ | Mat'ls £ | O & P £ | Total £ |
|---|---|---|---|---|---|---|
| Remove existing gutters and fittings, prepare to receive new gutterwork | | | | | | |
| cast iron | | | | | | |
| up to 100mm | nr | 0.50 | 7.25 | 0.00 | 1.09 | 8.34 |
| over 100mm | nr | 0.60 | 8.70 | 0.00 | 1.31 | 10.01 |
| PVC-U | | | | | | |
| up to 100mm | nr | 0.30 | 4.35 | 0.00 | 0.65 | 5.00 |
| over 100mm | nr | 0.35 | 5.08 | 0.00 | 0.76 | 5.84 |
| aluminium | | | | | | |
| up to 100mm | nr | 0.30 | 4.35 | 0.00 | 0.65 | 5.00 |
| over 100mm | nr | 0.35 | 5.08 | 0.00 | 0.76 | 5.84 |
| asbestos cement | | | | | | |
| up to 100mm | nr | 0.30 | 4.35 | 0.00 | 0.65 | 5.00 |
| over 100mm | nr | 0.35 | 5.08 | 0.00 | 0.76 | 5.84 |
| Remove single length of gutter, prepare ends and fix new length to existing brackets | | | | | | |
| cast iron | | | | | | |
| 100mm | nr | 2.00 | 29.00 | 21.70 | 7.61 | 58.31 |
| 115mm | nr | 2.10 | 30.45 | 22.40 | 7.93 | 60.78 |
| 125mm | nr | 2.15 | 31.18 | 25.80 | 8.55 | 65.52 |
| 150mm | nr | 2.30 | 33.35 | 43.28 | 11.49 | 88.12 |
| PVC-U | | | | | | |
| 76mm | nr | 1.00 | 14.50 | 12.14 | 4.00 | 30.64 |
| 112mm | nr | 1.05 | 15.23 | 9.96 | 3.78 | 28.96 |
| 150mm | nr | 1.10 | 15.95 | 13.21 | 4.37 | 33.53 |

| | Unit | Hours | Hours £ | Mat'ls £ | O & P £ | Total £ |
|---|---|---|---|---|---|---|
| **Remove single length of gutter (cont'd)** | | | | | | |
| aluminium | | | | | | |
| 100mm | nr | 1.90 | 27.55 | 24.28 | 7.77 | 59.60 |
| 125mm | nr | 2.05 | 29.73 | 30.98 | 9.11 | 69.81 |
| Remove gutter fittings, prepare ends and fix new fitting | | | | | | |
| cast iron, 100mm | | | | | | |
| angle | nr | 0.75 | 10.88 | 7.16 | 2.71 | 20.74 |
| stop end outlet | nr | 0.50 | 7.25 | 8.18 | 2.31 | 17.74 |
| cast iron, 115mm | | | | | | |
| angle | nr | 0.75 | 10.88 | 7.37 | 2.74 | 20.98 |
| stop end outlet | nr | 0.50 | 7.25 | 9.17 | 2.46 | 18.88 |
| cast iron, 125mm | | | | | | |
| angle | nr | 0.75 | 10.88 | 8.69 | 2.93 | 22.50 |
| stop end outlet | nr | 0.50 | 7.25 | 10.46 | 2.66 | 20.37 |
| cast iron, 150mm | | | | | | |
| angle | nr | 0.75 | 10.88 | 15.87 | 4.01 | 30.76 |
| stop end outlet | nr | 0.50 | 7.25 | 17.20 | 3.67 | 28.12 |
| PVC-U, 76mm | | | | | | |
| angle | nr | 0.50 | 7.25 | 6.03 | 1.99 | 15.27 |
| outlet | nr | 0.50 | 7.25 | 5.53 | 1.92 | 14.70 |
| stop end | nr | 0.50 | 7.25 | 2.97 | 1.53 | 11.75 |
| PVC-U, 112mm | | | | | | |
| angle | nr | 0.50 | 7.25 | 4.45 | 1.76 | 13.46 |
| outlet | nr | 0.50 | 7.25 | 3.90 | 1.67 | 12.82 |
| stop end | nr | 0.50 | 7.25 | 1.95 | 1.38 | 10.58 |

| | Unit | Hours | Hours £ | Mat'ls £ | O & P £ | Total £ |
|---|---|---|---|---|---|---|
| PVC-U, 150mm | | | | | | |
| angle | nr | 0.55 | 7.98 | 9.27 | 2.59 | 19.83 |
| outlet | nr | 0.55 | 7.98 | 5.56 | 2.03 | 15.57 |
| stop end | nr | 0.55 | 7.98 | 1.82 | 1.47 | 11.26 |
| aluminium, 100mm | | | | | | |
| angle | nr | 0.60 | 8.70 | 6.45 | 2.27 | 17.42 |
| outlet | nr | 0.60 | 8.70 | 10.96 | 2.95 | 22.61 |
| stop end | nr | 0.60 | 8.70 | 2.43 | 1.67 | 12.80 |
| aluminium, 160mm | | | | | | |
| angle | nr | 0.60 | 8.70 | 8.79 | 2.62 | 20.11 |
| outlet | nr | 0.60 | 8.70 | 7.25 | 2.39 | 18.34 |
| stop end | nr | 0.60 | 8.70 | 3.16 | 1.78 | 13.64 |
| Remove existing gutter brackets and replace with galvanised steel repair brackets at 1m maximum centres | | | | | | |
| cast iron | | | | | | |
| 100mm | nr | 0.25 | 3.63 | 1.91 | 0.83 | 6.37 |
| 112mm | nr | 0.27 | 3.92 | 1.91 | 0.87 | 6.70 |
| 125mm | nr | 0.30 | 4.35 | 1.91 | 0.94 | 7.20 |
| PVC-U | | | | | | |
| 76mm | nr | 0.25 | 3.63 | 0.71 | 0.65 | 4.99 |
| 112mm | nr | 0.27 | 3.92 | 1.22 | 0.77 | 5.91 |
| 150mm | nr | 0.30 | 4.35 | 1.38 | 0.86 | 6.59 |
| aluminium | | | | | | |
| 100mm | nr | 0.20 | 2.90 | 1.75 | 0.70 | 5.35 |
| 112mm | nr | 0.20 | 2.90 | 1.80 | 0.71 | 5.41 |
| 125mm | nr | 0.20 | 2.90 | 3.24 | 0.92 | 7.06 |

| | Unit | Hours | Hours £ | Mat'ls £ | O & P £ | Total £ |
|---|---|---|---|---|---|---|
| Remove pipes and fittings, prepare to receive new pipework | | | | | | |
| cast iron | | | | | | |
| up to 100mm | m | 0.50 | 7.25 | 0.00 | 1.09 | 8.34 |
| over 100mm | m | 0.60 | 8.70 | 0.00 | 1.31 | 10.01 |
| PVC-U | | | | | | |
| up to 100mm | m | 0.30 | 4.35 | 0.00 | 0.65 | 5.00 |
| over 100mm | m | 0.35 | 5.08 | 0.00 | 0.76 | 5.84 |
| aluminium | | | | | | |
| up to 100mm | m | 0.30 | 4.35 | 0.00 | 0.65 | 5.00 |
| over 100mm | m | 0.35 | 5.08 | 0.00 | 0.76 | 5.84 |
| Remove length of existing pipe, replace with new | | | | | | |
| cast iron | | | | | | |
| 65mm diameter | nr | 0.70 | 10.15 | 14.94 | 3.76 | 28.85 |
| 75mm diameter | nr | 0.70 | 10.15 | 41.02 | 7.68 | 58.85 |
| 100mm diameter | nr | 0.80 | 11.60 | 53.60 | 9.78 | 74.98 |
| PVC-U | | | | | | |
| 68mm diameter | nr | 0.50 | 7.25 | 11.22 | 2.77 | 21.24 |
| 68mm square | nr | 0.60 | 8.70 | 13.44 | 3.32 | 25.46 |
| aluminium | | | | | | |
| 63mm diameter | nr | 0.50 | 7.25 | 30.66 | 5.69 | 43.60 |
| 76mm diameter | nr | 0.50 | 7.25 | 37.78 | 6.75 | 51.78 |
| 102mm diameter | nr | 0.60 | 8.70 | 52.76 | 9.22 | 70.68 |
| Remove existing pipe fittings, replace with new | | | | | | |
| cast iron offset | | | | | | |
| 65mm diameter | nr | 0.28 | 4.06 | 15.28 | 2.90 | 22.24 |
| 75mm diameter | nr | 0.30 | 4.35 | 15.28 | 2.94 | 22.57 |
| 100mm diameter | nr | 0.32 | 4.64 | 29.39 | 5.10 | 39.13 |

| | Unit | Hours | Hours £ | Mat'ls £ | O & P £ | Total £ |
|---|---|---|---|---|---|---|
| cast iron shoe | | | | | | |
| 65mm diameter | nr | 0.36 | 5.22 | 14.14 | 2.90 | 22.26 |
| 75mm diameter | nr | 0.38 | 5.51 | 21.26 | 4.02 | 30.79 |
| 100mm diameter | nr | 0.40 | 5.80 | 19.06 | 3.73 | 28.59 |
| cast iron shoe, eared | | | | | | |
| 65mm diameter | nr | 0.36 | 5.22 | 16.30 | 3.23 | 24.75 |
| 75mm diameter | nr | 0.38 | 5.51 | 16.30 | 3.27 | 25.08 |
| 100mm diameter | nr | 0.40 | 5.80 | 21.66 | 4.12 | 31.58 |
| cast iron bend | | | | | | |
| 65mm diameter | nr | 0.20 | 2.90 | 9.98 | 1.93 | 14.81 |
| 75mm diameter | nr | 0.22 | 3.19 | 12.12 | 2.30 | 17.61 |
| 100mm diameter | nr | 0.24 | 3.48 | 17.12 | 3.09 | 23.69 |
| PVC-U bend | | | | | | |
| 68mm diameter | nr | 0.40 | 5.80 | 5.39 | 1.68 | 12.87 |
| 68mm square | nr | 0.40 | 5.80 | 3.44 | 1.39 | 10.63 |
| PVC-U offset | | | | | | |
| 68mm diameter | nr | 0.20 | 2.90 | 3.21 | 0.92 | 7.03 |
| 68mm square | nr | 0.20 | 2.90 | 6.04 | 1.34 | 10.28 |
| PVC-U shoe | | | | | | |
| 68mm diameter | nr | 0.20 | 2.90 | 4.65 | 1.13 | 8.68 |
| 68mm square | nr | 0.20 | 2.90 | 3.95 | 1.03 | 7.88 |
| aluminium offset | | | | | | |
| 63mm diameter | nr | 0.30 | 4.35 | 29.01 | 5.00 | 38.36 |
| 76mm diameter | nr | 0.32 | 4.64 | 35.69 | 6.05 | 46.38 |
| 102mm diameter | nr | 0.24 | 3.48 | 43.39 | 7.03 | 53.90 |
| aluminium bend | | | | | | |
| 63mm diameter | nr | 0.38 | 5.51 | 23.61 | 4.37 | 33.49 |
| 76mm diameter | nr | 0.40 | 5.80 | 26.54 | 4.85 | 37.19 |
| 102mm diameter | nr | 0.42 | 6.09 | 38.05 | 6.62 | 50.76 |

| | Unit | Hours | Hours £ | Mat'ls £ | O & P £ | Total £ |
|---|---|---|---|---|---|---|
| **Remove pipe fittings and replace with new (cont'd)** | | | | | | |
| Remove existing gutter fittings, replace with new | | | | | | |
| cast iron angle | | | | | | |
| 100mm diameter | nr | 0.44 | 6.38 | 7.16 | 2.03 | 15.57 |
| 125mm diameter | nr | 0.46 | 6.67 | 8.69 | 2.30 | 17.66 |
| 150mm diameter | nr | 0.48 | 6.96 | 15.87 | 3.42 | 26.25 |
| cast iron stop end outlet | | | | | | |
| 100mm diameter | nr | 0.30 | 4.35 | 8.18 | 1.88 | 14.41 |
| 125mm diameter | nr | 0.32 | 4.64 | 10.46 | 2.27 | 17.37 |
| 150mm diameter | nr | 0.34 | 4.93 | 17.20 | 3.32 | 25.45 |
| PVC-U running outlet | | | | | | |
| 76mm | nr | 0.20 | 2.90 | 5.53 | 1.26 | 9.69 |
| 112mm | nr | 0.22 | 3.19 | 3.90 | 1.06 | 8.15 |
| 150mm | nr | 0.24 | 3.48 | 9.00 | 1.87 | 14.35 |
| PVC-U angle | | | | | | |
| 76mm | nr | 0.20 | 2.90 | 6.03 | 1.34 | 10.27 |
| 112mm | nr | 0.22 | 3.19 | 4.45 | 1.15 | 8.79 |
| 150mm | nr | 0.24 | 3.48 | 9.27 | 1.91 | 14.66 |
| PVC-U stop end outlet | | | | | | |
| 76mm | nr | 0.20 | 2.90 | 4.85 | 1.16 | 8.91 |
| 112mm | nr | 0.22 | 3.19 | 3.90 | 1.06 | 8.15 |
| 150mm | nr | 0.24 | 3.48 | 5.56 | 1.36 | 10.40 |
| aluminium running outlet | | | | | | |
| 100mm | nr | 0.38 | 5.51 | 10.96 | 2.47 | 18.94 |
| 125mm | nr | 0.42 | 6.09 | 10.48 | 2.49 | 19.06 |

| | Unit | Hours | Hours £ | Mat'ls £ | O & P £ | Total £ |
|---|---|---|---|---|---|---|
| aluminium angle | | | | | | |
| 100mm | nr | 0.38 | 5.51 | 6.45 | 1.79 | 13.75 |
| 125mm | nr | 0.42 | 6.09 | 8.79 | 2.23 | 17.11 |
| aluminium stop end outlet | | | | | | |
| 100mm | nr | 0.38 | 5.51 | 6.30 | 1.77 | 13.58 |
| 125mm | nr | 0.42 | 6.09 | 7.25 | 2.00 | 15.34 |
| Rake out existing cement mortar joint of soil pipe connection and re-point | | | | | | |
| 65mm diameter pipe | nr | 0.40 | 5.80 | 0.50 | 0.95 | 7.25 |
| 75mm diameter pipe | nr | 0.45 | 6.53 | 0.60 | 1.07 | 8.19 |
| 100mm diameter pipe | nr | 0.50 | 7.25 | 0.70 | 1.19 | 9.14 |
| **Hot and cold water supply and heating installation** | | | | | | |
| Cut out 500mm length of copper pipe, fix new length with brass compression connections to existing ends | | | | | | |
| 15mm | nr | 0.40 | 5.80 | 3.46 | 1.39 | 10.65 |
| 18mm | nr | 0.40 | 5.80 | 4.90 | 1.61 | 12.31 |
| 22mm | nr | 0.40 | 5.80 | 5.61 | 1.71 | 13.12 |
| 28mm | nr | 0.45 | 6.53 | 7.92 | 2.17 | 16.61 |
| 35mm | nr | 0.45 | 6.53 | 14.54 | 3.16 | 24.22 |
| 42mm | nr | 0.50 | 7.25 | 18.28 | 3.83 | 29.36 |
| 54mm | nr | 0.50 | 7.25 | 22.72 | 4.50 | 34.47 |
| Remove existing pipework and prepare to receive new | | | | | | |
| copper | | | | | | |
| up to 25mm | nr | 0.20 | 2.90 | 0.00 | 0.44 | 3.34 |
| 25-54mm | nr | 0.30 | 4.35 | 0.00 | 0.65 | 5.00 |

| | Unit | Hours | Hours £ | Mat'ls £ | O & P £ | Total £ |
|---|---|---|---|---|---|---|
| Remove existing pipework and prepare to receive new (cont'd) | | | | | | |
| lead | | | | | | |
| up to 1in | nr | 0.50 | 7.25 | 0.00 | 1.09 | 8.34 |
| Take off existing radiator valve, replace with new, drain down system prior to removal and bleed after installation | | | | | | |
| single valves, standard type | nr | 0.40 | 5.80 | 16.94 | 3.41 | 26.15 |
| complete systems, 9nr valves, standard type | nr | 4.00 | 58.00 | 152.46 | 31.57 | 242.03 |
| Take out existing galvanised steel water storage tank, fix new plastic tank complete with ball valve, lid and insulation, allow for cutting holes | | | | | | |
| 18 litres tank | nr | 3.00 | 43.50 | 18.32 | 9.27 | 71.09 |
| 68 litres tank | nr | 3.50 | 50.75 | 27.51 | 11.74 | 90.00 |
| 154 litres tank | nr | 4.30 | 62.35 | 49.59 | 16.79 | 128.73 |
| **Sanitary fittings** | | | | | | |
| Take out existing cast iron bath including trap and taps, cut back pipework as necessary and fix new acrylic reinforced bath complete with trap, bath panels, taps and shower handset and connect to existing pipework | nr | 4.00 | 58.00 | 342.51 | 60.08 | 460.59 |

| | Unit | Hours | Hours £ | Mat'ls £ | O & P £ | Total £ |
|---|---|---|---|---|---|---|
| Take out existing wash basin including trap and taps, cut back pipework as necessary and fix new pedestal-mounted wash basin complete with trap and taps and connect to existing pipework | nr | 3.00 | 43.50 | 112.48 | 23.40 | 179.38 |
| Take out existing high-level WC, cut back pipework as necessary and fix new low-level WC suite complete and connect to existing pipework | nr | 3.50 | 50.75 | 272.85 | 48.54 | 372.14 |

# Part Two

## PROJECT COSTS

Rainwater goods

Bathrooms

External waste systems

Central heating systems

Hot and cold water supply systems

| | Qty | Unit | Hours | Hours £ | Mat'ls £ | O & P £ | Total £ |
|---|---|---|---|---|---|---|---|
| **PVC-U RAINWATER GOODS** | | | | | | | |
| **One-storey terraced house size 7 x 6m with gable ends** | | | | | | | |
| Take down existing rainwater goods and make good | | Item | 4.00 | 58.00 | 0.00 | 8.70 | 66.70 |
| 68mm diameter PVC-U rainwater pipe with pipe clips plugged to brickwork at 2m maximum centres | 6 | m | 1.50 | 21.75 | 33.66 | 8.31 | 63.72 |
| extra over for | | | | | | | |
| branch | 2 | nr | 0.50 | 7.25 | 6.21 | 2.02 | 15.48 |
| shoe | 2 | nr | 0.50 | 7.25 | 9.30 | 2.48 | 19.03 |
| 110mm half round PVC-U gutter to softwood fascia | 14 | m | 2.80 | 40.60 | 62.72 | 15.50 | 118.82 |
| extra over for | | | | | | | |
| running outlet | 2 | nr | 0.52 | 7.54 | 7.80 | 2.30 | 17.64 |
| **Total** | | | 9.82 | 142.39 | 119.69 | 39.31 | 301.39 |
| **One-storey semi-detached house size 8 x 7m with gable ends** | | | | | | | |
| Take down existing rainwater goods and make good | | Item | 6.00 | 87.00 | 0.00 | 13.05 | 100.05 |
| 68mm diameter PVC-U rainwater pipe with pipe clips plugged to brickwork at 2m maximum centres | 6 | m | 1.50 | 21.75 | 33.66 | 8.31 | 63.72 |
| **Carried forward** | | | 7.50 | 108.75 | 33.66 | 21.36 | 163.77 |

| | Qty | Unit | Hours | Hours £ | Mat'ls £ | O & P £ | Total £ |
|---|---|---|---|---|---|---|---|
| **Brought forward** | | | 7.50 | 108.75 | 33.66 | 21.36 | 163.77 |
| extra over for | | | | | | | |
| offset | 2 | nr | 0.30 | 4.35 | 6.22 | 1.59 | 12.16 |
| shoe | 2 | nr | 0.60 | 8.70 | 9.30 | 2.70 | 20.70 |
| 110mm half round PVC-U gutter to softwood fascia | 16 | m | 3.20 | 46.40 | 71.68 | 17.71 | 135.79 |
| extra over for | | | | | | | |
| stop end outlet | 2 | nr | 0.26 | 3.77 | 3.90 | 1.15 | 8.82 |
| **Total** | | | 11.86 | 171.97 | 124.76 | 44.51 | 341.24 |

| | Qty | Unit | Hours | Hours £ | Mat'ls £ | O & P £ | Total £ |
|---|---|---|---|---|---|---|---|
| **One-storey detached house size 9 x 8m with gable ends** | | | | | | | |
| Take down existing rainwater goods and make good | | Item | 7.00 | 101.50 | 0.00 | 15.23 | 116.73 |
| 68mm diameter PVC-U rainwater pipe with pipe clips plugged to brickwork at 2m maximum centres | 6 | m | 1.50 | 21.75 | 33.66 | 8.31 | 63.72 |
| extra over for | | | | | | | |
| offset | 2 | nr | 0.30 | 4.35 | 6.22 | 1.59 | 12.16 |
| shoe | 2 | nr | 0.60 | 8.70 | 9.30 | 2.70 | 20.70 |
| 110mm half round PVC-U gutter to softwood fascia | 18 | m | 3.60 | 52.20 | 80.64 | 19.93 | 152.77 |
| extra over for | | | | | | | |
| running outlet | 2 | nr | 0.26 | 3.77 | 3.90 | 1.15 | 8.82 |
| **Total** | | | 13.26 | 192.27 | 133.72 | 48.90 | 374.89 |

| | Qty | Unit | Hours | Hours £ | Mat'ls £ | O & P £ | Total £ |
|---|---|---|---|---|---|---|---|
| **Two-storey terraced house size 7 x 7m with gable ends** | | | | | | | |
| Take down existing rainwater goods and make good | | Item | 5.00 | 72.50 | 0.00 | 10.88 | 83.38 |
| 68mm diameter PVC-U rainwater pipe with pipe clips plugged to brickwork at 2m maximum centres | 12 | m | 3.00 | 43.50 | 67.32 | 16.62 | 127.44 |
| extra over for | | | | | | | |
| offset | 2 | nr | 0.50 | 7.25 | 6.42 | 2.05 | 15.72 |
| shoe | 2 | nr | 0.50 | 7.25 | 9.30 | 2.48 | 19.03 |
| 110mm half round PVC-U gutter to softwood fascia | 14 | m | 2.80 | 40.60 | 71.68 | 16.84 | 129.12 |
| extra over for | | | | | | | |
| running outlet | 2 | nr | 0.52 | 7.54 | 3.90 | 1.72 | 13.16 |
| **Total** | | | 12.32 | 178.64 | 158.62 | 50.59 | 387.85 |

| | Qty | Unit | Hours | Hours £ | Mat'ls £ | O & P £ | Total £ |
|---|---|---|---|---|---|---|---|
| **Two-storey semi-detached house size 8 x 7m with gable ends** | | | | | | | |
| Take down existing rainwater goods and make good | | Item | 7.00 | 101.50 | 0.00 | 15.23 | 116.73 |
| 68mm diameter PVC-U rainwater pipe with pipe clips plugged to brickwork at 2m maximum centres | 12 | m | 3.00 | 43.50 | 67.32 | 16.62 | 127.44 |
| **Carried forward** | | | 10.00 | 145.00 | 67.32 | 31.85 | 244.17 |

| | Qty | Unit | Hours | Hours £ | Mat'ls £ | O & P £ | Total £ |
|---|---|---|---|---|---|---|---|
| **Brought forward** | | | 10.00 | 145.00 | 67.32 | 31.85 | 244.17 |
| extra over for | | | | | | | |
| offset | 2 | nr | 0.50 | 7.25 | 6.42 | 2.05 | 15.72 |
| shoe | 2 | nr | 0.50 | 7.25 | 9.30 | 2.48 | 19.03 |
| 110mm half round PVC-U gutter to softwood fascia | 16 | m | 3.60 | 52.20 | 80.64 | 19.93 | 152.77 |
| extra over for | | | | | | | |
| stop end outlet | 2 | nr | 0.26 | 3.77 | 3.90 | 1.15 | 8.82 |
| **Total** | | | 14.86 | 215.47 | 167.58 | 57.46 | 440.51 |

| | Qty | Unit | Hours | Hours £ | Mat'ls £ | O & P £ | Total £ |
|---|---|---|---|---|---|---|---|
| **Two-storey detached house size 9 x 8m with gable ends** | | | | | | | |
| Take down existing rainwater goods and make good | | Item | 8.00 | 116.00 | 0.00 | 17.40 | 133.40 |
| 68mm diameter PVC-U rainwater pipe with pipe clips plugged to brickwork at 2m maximum centres | 12 | m | 3.00 | 43.50 | 67.32 | 16.62 | 127.44 |
| extra over for | | | | | | | |
| offset | 2 | nr | 0.50 | 7.25 | 6.43 | 2.05 | 15.73 |
| shoe | 2 | nr | 0.50 | 7.25 | 9.30 | 2.48 | 19.03 |
| 110mm half round PVC-U gutter to softwood fascia | 18 | m | 3.60 | 52.20 | 80.64 | 19.93 | 152.77 |
| extra over for | | | | | | | |
| stop end outlet | 2 | nr | 0.26 | 3.77 | 3.90 | 1.15 | 8.82 |
| **Total** | | | 15.86 | 229.97 | 167.59 | 59.63 | 457.19 |

| | Qty | Unit | Hours | Hours £ | Mat'ls £ | O & P £ | Total £ |
|---|---|---|---|---|---|---|---|
| **Three-storey terraced house size 8 x 8m with gable ends** | | | | | | | |
| Take down existing rainwater goods and make good | | Item | 6.00 | 87.00 | 0.00 | 13.05 | 100.05 |
| 68mm diameter PVC-U rainwater pipe with pipe clips plugged to brickwork at 2m maximum centres | 18 | m | 4.50 | 65.25 | 100.98 | 24.93 | 191.16 |
| extra over for | | | | | | | |
| offset | 2 | nr | 0.50 | 7.25 | 6.42 | 2.05 | 15.72 |
| shoe | 2 | nr | 0.50 | 7.25 | 9.30 | 2.48 | 19.03 |
| 110mm half round PVC-U gutter to softwood fascia | 16 | m | 3.20 | 46.40 | 71.68 | 17.71 | 135.79 |
| extra over for | | | | | | | |
| running outlet | 2 | nr | 0.52 | 7.54 | 7.80 | 2.30 | 17.64 |
| **Total** | | | 15.22 | 220.69 | 196.18 | 62.53 | 479.40 |

| | Qty | Unit | Hours | Hours £ | Mat'ls £ | O & P £ | Total £ |
|---|---|---|---|---|---|---|---|
| **Three-storey semi-detached house size 9 x 8m with gable ends** | | | | | | | |
| Take down existing rainwater goods and make good | | Item | 7.00 | 101.50 | 0.00 | 15.23 | 116.73 |
| 68mm diameter PVC-U rainwater pipe with pipe clips plugged to brickwork at 2m maximum centres | 18 | m | 4.50 | 65.25 | 100.98 | 24.93 | 191.16 |
| **Carried forward** | | | 11.50 | 166.75 | 100.98 | 40.16 | 307.89 |

| | Qty | Unit | Hours | Hours £ | Mat'ls £ | O & P £ | Total £ |
|---|---|---|---|---|---|---|---|
| **Brought forward** | | | 11.50 | 166.75 | 100.98 | 40.16 | 307.89 |
| extra over for | | | | | | | |
| offset | 2 | nr | 0.50 | 7.25 | 6.42 | 2.05 | 15.72 |
| shoe | 2 | nr | 0.50 | 7.25 | 9.30 | 2.48 | 19.03 |
| 110mm half round PVC-U gutter to softwood fascia | 18 | m | 3.60 | 52.20 | 80.64 | 19.93 | 152.77 |
| extra over for | | | | | | | |
| stop end outlet | 2 | nr | 0.26 | 3.77 | 3.90 | 1.15 | 8.82 |
| **Total** | | | 16.36 | 237.22 | 201.24 | 65.77 | 504.23 |

| | Qty | Unit | Hours | Hours £ | Mat'ls £ | O & P £ | Total £ |
|---|---|---|---|---|---|---|---|
| **Three-storey detached house size 10 x 9m with gable ends** | | | | | | | |
| Take down existing rainwater goods and make good | | Item | 9.00 | 130.50 | 0.00 | 19.58 | 150.08 |
| 68mm diameter PVC-U rainwater pipe with pipe clips plugged to brickwork at 2m maximum centres | 18 | m | 4.50 | 65.25 | 100.98 | 24.93 | 191.16 |
| extra over for | | | | | | | |
| offset | 2 | nr | 0.50 | 7.25 | 6.42 | 2.05 | 15.72 |
| shoe | 2 | nr | 0.50 | 7.25 | 9.30 | 2.48 | 19.03 |
| 110mm half round PVC-U gutter to softwood fascia | 20 | m | 4.00 | 58.00 | 89.60 | 22.14 | 169.74 |
| extra over for | | | | | | | |
| stop end outlet | 2 | nr | 0.26 | 3.77 | 3.90 | 1.15 | 8.82 |
| **Total** | | | 18.76 | 272.02 | 210.20 | 72.33 | 554.55 |

| | Qty | Unit | Hours | Hours £ | Mat'ls £ | O & P £ | Total £ |
|---|---|---|---|---|---|---|---|
| **One-storey terraced house size 7 x 6m with hipped ends** | | | | | | | |
| Take down existing rainwater goods and make good | | Item | 4.00 | 58.00 | 0.00 | 8.70 | 66.70 |
| 68mm diameter PVC-U rainwater pipe with pipe clips plugged to brickwork at 2m maximum centres | 6 | m | 1.50 | 21.75 | 33.66 | 8.31 | 63.72 |
| extra over for | | | | | | | |
| offset | 2 | nr | 0.50 | 7.25 | 6.42 | 2.05 | 15.72 |
| shoe | 2 | nr | 0.50 | 7.25 | 9.30 | 2.48 | 19.03 |
| 110mm half round PVC-U gutter to softwood fascia | 14 | m | 2.80 | 40.60 | 62.72 | 15.50 | 118.82 |
| extra over for | | | | | | | |
| running outlet | 2 | nr | 0.52 | 7.54 | 7.80 | 2.30 | 17.64 |
| **Total** | | | 9.82 | 142.39 | 119.90 | 39.34 | 301.63 |
| **One-storey semi-detached house size 8 x 7m with hipped ends** | | | | | | | |
| Take down existing rainwater goods and make good | | Item | 6.00 | 87.00 | 0.00 | 13.05 | 100.05 |
| 68mm diameter PVC-U rainwater pipe with pipe clips plugged to brickwork at 2m maximum centres | 6 | m | 1.50 | 21.75 | 33.66 | 8.31 | 63.72 |
| extra over for | | | | | | | |
| offset | 2 | nr | 0.50 | 7.25 | 6.42 | 2.05 | 15.72 |
| shoe | 2 | nr | 0.50 | 7.25 | 9.30 | 2.48 | 19.03 |
| **Carried forward** | | | 8.50 | 123.25 | 49.38 | 25.89 | 198.52 |

| | Qty | Unit | Hours | Hours £ | Mat'ls £ | O & P £ | Total £ |
|---|---|---|---|---|---|---|---|
| **Brought forward** | | | 8.50 | 123.25 | 49.38 | 25.89 | 198.52 |
| 110mm half round PVC-U gutter to softwood fascia | 23 | m | 4.60 | 66.70 | 103.04 | 25.46 | 195.20 |
| extra over for | | | | | | | |
| angle | 2 | nr | 0.52 | 7.54 | 0.52 | 1.21 | 9.27 |
| running outlet | 2 | nr | 0.52 | 7.54 | 0.52 | 1.21 | 9.27 |
| **Total** | | | 14.14 | 205.03 | 153.46 | 53.77 | 412.26 |

| | Qty | Unit | Hours | Hours £ | Mat'ls £ | O & P £ | Total £ |
|---|---|---|---|---|---|---|---|
| **One-storey detached house size 9 x 7m with hipped ends** | | | | | | | |
| Take down existing rainwater goods and make good | | Item | 8.00 | 116.00 | 0.00 | 17.40 | 133.40 |
| 68mm diameter PVC-U rainwater pipe with pipe clips plugged to brickwork at 2m maximum centres | 6 | m | 1.50 | 21.75 | 33.66 | 8.31 | 63.72 |
| extra over for | | | | | | | |
| offset | 2 | nr | 0.50 | 7.25 | 6.42 | 2.05 | 15.72 |
| shoe | 2 | nr | 0.50 | 7.25 | 9.30 | 2.48 | 19.03 |
| 110mm half round PVC-U gutter to softwood fascia | 32 | m | 6.40 | 92.80 | 143.46 | 35.44 | 271.70 |
| extra over for | | | | | | | |
| angle | 4 | nr | 0.52 | 7.54 | 7.80 | 2.30 | 17.64 |
| running outlet | 2 | nr | 1.04 | 15.08 | 17.80 | 4.93 | 37.81 |
| **Total** | | | 18.46 | 267.67 | 218.44 | 72.92 | 559.03 |

| | Qty | Unit | Hours | Hours £ | Mat'ls £ | O & P £ | Total £ |
|---|---|---|---|---|---|---|---|
| **Two-storey terraced house size 7 x 7m with hipped ends** | | | | | | | |
| Take down existing rainwater goods and make good | | Item | 6.00 | 87.00 | 0.00 | 13.05 | 100.05 |
| 68mm diameter PVC-U rainwater pipe with pipe clips plugged to brickwork at 2m maximum centres | 12 | m | 3.00 | 43.50 | 67.32 | 16.62 | 127.44 |
| extra over for | | | | | | | |
| offset | 2 | nr | 0.50 | 7.25 | 6.42 | 2.05 | 15.72 |
| shoe | 2 | nr | 0.50 | 7.25 | 9.30 | 2.48 | 19.03 |
| 110mm half round PVC-U gutter to softwood fascia | 14 | m | 2.80 | 40.60 | 62.72 | 15.50 | 118.82 |
| extra over for | | | | | | | |
| running outlet | 2 | nr | 0.52 | 7.54 | 7.80 | 2.30 | 17.64 |
| **Total** | | | 13.32 | 193.14 | 153.56 | 52.01 | 398.71 |

| | Qty | Unit | Hours | Hours £ | Mat'ls £ | O & P £ | Total £ |
|---|---|---|---|---|---|---|---|
| **Two-storey semi-detached house size 8 x 7m with hipped ends** | | | | | | | |
| Take down existing rainwater goods and make good | | Item | 8.00 | 116.00 | 0.00 | 17.40 | 133.40 |
| 68mm diameter PVC-U rainwater pipe with pipe clips plugged to brickwork at 2m maximum centres | 12 | m | 3.00 | 43.50 | 67.32 | 16.62 | 127.44 |
| **Carried forward** | | | 11.00 | 159.50 | 67.32 | 34.02 | 260.84 |

| | Qty | Unit | Hours | Hours £ | Mat'ls £ | O & P £ | Total £ |
|---|---|---|---|---|---|---|---|
| **Brought forward** | | | 11.00 | 159.50 | 67.32 | 34.02 | 260.84 |
| extra over for | | | | | | | |
| offset | 2 | nr | 0.50 | 7.25 | 6.42 | 2.05 | 15.72 |
| shoe | 2 | nr | 0.50 | 7.25 | 9.30 | 2.48 | 19.03 |
| 110mm half round PVC-U gutter to softwood fascia | 24 | m | 4.60 | 66.70 | 103.04 | 25.46 | 195.20 |
| extra over for | | | | | | | |
| angle | 2 | nr | 0.52 | 7.54 | 7.80 | 2.30 | 17.64 |
| running outlet | 2 | nr | 0.52 | 7.54 | 9.90 | 2.62 | 20.06 |
| **Total** | | | 17.64 | 255.78 | 203.78 | 68.93 | 528.49 |

| | Qty | Unit | Hours | Hours £ | Mat'ls £ | O & P £ | Total £ |
|---|---|---|---|---|---|---|---|
| **Two-storey detached house size 9 x 8m with hipped ends** | | | | | | | |
| Take down existing rainwater goods and make good | | Item | 9.00 | 130.50 | 0.00 | 19.58 | 150.08 |
| 68mm diameter PVC-U rainwater pipe as before | 12 | m | 3.00 | 43.50 | 67.32 | 16.62 | 127.44 |
| extra over for | | | | | | | |
| offset | 2 | nr | 0.50 | 7.25 | 6.42 | 2.05 | 15.72 |
| shoe | 2 | nr | 0.50 | 7.25 | 9.30 | 2.48 | 19.03 |
| 110mm half round PVC-U gutter to softwood fascia | 34 | m | 6.80 | 98.60 | 152.32 | 37.64 | 288.56 |
| extra over for | | | | | | | |
| angle | 4 | nr | 0.52 | 7.54 | 7.80 | 2.30 | 17.64 |
| running outlet | 2 | nr | 1.04 | 15.08 | 17.80 | 4.93 | 37.81 |
| **Total** | | | 21.36 | 309.72 | 260.96 | 85.60 | 656.28 |

| | Qty | Unit | Hours | Hours £ | Mat'ls £ | O & P £ | Total £ |
|---|---|---|---|---|---|---|---|
| **Three-storey terraced house size 8 x 8m with hipped ends** | | | | | | | |
| Take down existing rainwater goods and make good | | Item | 7.00 | 101.50 | 0.00 | 15.23 | 116.73 |
| 68mm diameter PVC-U rainwater pipe with pipe clips plugged to brickwork at 2m maximum centres | 18 | m | 4.50 | 65.25 | 100.98 | 24.93 | 191.16 |
| extra over for | | | | | | | |
| offset | 2 | nr | 0.50 | 7.25 | 6.42 | 2.05 | 15.72 |
| shoe | 2 | nr | 0.50 | 7.25 | 9.30 | 2.48 | 19.03 |
| 110mm half round PVC-U gutter to softwood fascia | 16 | m | 3.20 | 46.40 | 71.68 | 17.71 | 135.79 |
| extra over for | | | | | | | |
| running outlet | 2 | nr | 0.52 | 7.54 | 7.80 | 2.30 | 17.64 |
| **Total** | | | 16.22 | 235.19 | 196.18 | 64.71 | 496.08 |

| | Qty | Unit | Hours | Hours £ | Mat'ls £ | O & P £ | Total £ |
|---|---|---|---|---|---|---|---|
| **Three-storey semi-detached house size 9 x 8m with hipped ends** | | | | | | | |
| Take down existing rainwater goods and make good | | Item | 9.00 | 130.50 | 0.00 | 19.58 | 150.08 |
| 68mm diameter PVC-U rainwater pipe with pipe clips plugged to brickwork at 2m maximum centres | 18 | m | 4.50 | 65.25 | 100.98 | 24.93 | 191.16 |
| **Carried forward** | | | 13.50 | 195.75 | 100.98 | 44.51 | 341.24 |

| | Qty | Unit | Hours | Hours £ | Mat'ls £ | O & P £ | Total £ |
|---|---|---|---|---|---|---|---|
| **Brought forward** | | | 13.50 | 195.75 | 100.98 | 44.51 | 341.24 |
| extra over for | | | | | | | |
| offset | 2 | nr | 0.50 | 7.25 | 6.42 | 2.05 | 15.72 |
| shoe | 2 | nr | 0.50 | 7.25 | 9.30 | 2.48 | 19.03 |
| 110mm half round PVC-U gutter to softwood fascia | 26 | m | 5.20 | 75.40 | 116.48 | 28.78 | 220.66 |
| extra over for | | | | | | | |
| angle | 2 | nr | 0.52 | 7.54 | 7.80 | 2.30 | 17.64 |
| running outlet | 2 | nr | 0.52 | 7.54 | 7.80 | 2.30 | 17.64 |
| **Total** | | | 20.74 | 300.73 | 248.78 | 82.43 | 631.94 |

| | Qty | Unit | Hours | Hours £ | Mat'ls £ | O & P £ | Total £ |
|---|---|---|---|---|---|---|---|
| **Three-storey detached house size 10 x 9m with hipped ends** | | | | | | | |
| Take down existing rainwater goods and make good | | Item | 10.00 | 145.00 | 0.00 | 21.75 | 166.75 |
| 68mm diameter PVC-U rainwater pipe as before | 18 | m | 4.50 | 65.25 | 100.98 | 24.93 | 191.16 |
| extra over for | | | | | | | |
| offset | 2 | nr | 0.50 | 7.25 | 6.42 | 2.05 | 15.72 |
| shoe | 2 | nr | 0.50 | 7.25 | 9.30 | 2.48 | 19.03 |
| 110mm half round PVC-U gutter to softwood fascia | 38 | m | 7.60 | 110.20 | 170.24 | 42.07 | 322.51 |
| extra over for | | | | | | | |
| angle | 4 | nr | 0.52 | 7.54 | 7.80 | 2.30 | 17.64 |
| running outlet | 2 | nr | 0.52 | 7.54 | 7.80 | 2.30 | 17.64 |
| **Total** | | | 24.14 | 350.03 | 302.54 | 97.89 | 750.46 |

| | Qty | Unit | Hours | Hours £ | Mat'ls £ | O & P £ | Total £ |
|---|---|---|---|---|---|---|---|
| **CAST IRON RAINWATER GOODS** | | | | | | | |
| **One-storey terraced house size 7 x 6m with gable ends** | | | | | | | |
| Take down existing rainwater goods and make good | | Item | 4.00 | 58.00 | 0.00 | 8.70 | 66.70 |
| 75mm diameter cast iron rainwater pipe with pipe clips plugged to brickwork at 2m maximum centres | 6 | m | 1.80 | 26.10 | 123.06 | 22.37 | 171.53 |
| extra over for | | | | | | | |
| offset | 2 | nr | 0.60 | 8.70 | 30.56 | 5.89 | 45.15 |
| shoe | 2 | nr | 0.60 | 8.70 | 28.28 | 5.55 | 42.53 |
| 100mm half round cast iron gutter to softwood fascia | 14 | m | 4.20 | 60.90 | 172.06 | 34.94 | 267.90 |
| extra over for | | | | | | | |
| running outlet | 2 | nr | 0.60 | 8.70 | 21.92 | 4.59 | 35.21 |
| **Total** | | | 11.80 | 171.10 | 375.88 | 82.05 | 629.03 |
| **One-storey semi-detached house size 8 x 7m with gable ends** | | | | | | | |
| Take down existing rainwater goods and make good | | Item | 6.00 | 87.00 | 0.00 | 13.05 | 100.05 |
| 75mm diameter cast iron rainwater pipe with pipe clips plugged to brickwork at 2m maximum centres | 6 | m | 1.80 | 26.10 | 123.06 | 22.37 | 171.53 |
| **Carried forward** | | | 7.80 | 113.10 | 123.06 | 35.42 | 271.58 |

| | Qty | Unit | Hours | Hours £ | Mat'ls £ | O & P £ | Total £ |
|---|---|---|---|---|---|---|---|
| **Brought forward** | | | 7.80 | 113.10 | 123.06 | 35.42 | 271.58 |
| extra over for | | | | | | | |
| offset | 2 | nr | 0.60 | 8.70 | 30.56 | 5.89 | 45.15 |
| shoe | 2 | nr | 0.60 | 8.70 | 28.28 | 5.55 | 42.53 |
| 100mm half round cast iron gutter to softwood fascia | 16 | m | 4.80 | 69.60 | 196.64 | 39.94 | 306.18 |
| extra over for | | | | | | | |
| stop end outlet | 2 | nr | 0.30 | 4.35 | 12.60 | 2.54 | 19.49 |
| **Total** | | | 14.10 | 204.45 | 391.14 | 89.33 | 684.92 |

| | Qty | Unit | Hours | Hours £ | Mat'ls £ | O & P £ | Total £ |
|---|---|---|---|---|---|---|---|
| **One-storey detached house size 9 x 8m with gable ends** | | | | | | | |
| Take down existing rainwater goods and make good | | Item | 7.00 | 101.50 | 0.00 | 15.23 | 116.73 |
| 75mm diameter cast iron rainwater pipe with pipe clips plugged to brickwork at 2m maximum centres | 6 | m | 1.80 | 26.10 | 123.06 | 22.37 | 171.53 |
| extra over for | | | | | | | |
| offset | 2 | nr | 0.60 | 8.70 | 30.56 | 5.89 | 45.15 |
| shoe | 2 | nr | 0.60 | 8.70 | 28.28 | 5.55 | 42.53 |
| 100mm half round cast iron gutter to softwood fascia | 18 | m | 5.40 | 78.30 | 221.22 | 44.93 | 344.45 |
| extra over for | | | | | | | |
| running outlet | 2 | nr | 0.30 | 4.35 | 12.60 | 2.54 | 19.49 |
| **Total** | | | 15.70 | 227.65 | 415.72 | 96.51 | 739.88 |

| | Qty | Unit | Hours | Hours £ | Mat'ls £ | O & P £ | Total £ |
|---|---|---|---|---|---|---|---|
| **Two-storey terraced house size 7 x 7m with gable ends** | | | | | | | |
| Take down existing rainwater goods and make good | | Item | 5.00 | 72.50 | 0.00 | 10.88 | 83.38 |
| 75mm diameter cast iron rainwater pipe with pipe clips plugged to brickwork at 2m maximum centres | 12 | m | 3.60 | 52.20 | 246.12 | 44.75 | 343.07 |
| extra over for | | | | | | | |
| offset | 2 | nr | 0.60 | 8.70 | 30.56 | 5.89 | 45.15 |
| shoe | 2 | nr | 0.60 | 8.70 | 28.28 | 5.55 | 42.53 |
| 100mm half round cast iron gutter to softwood fascia | 14 | m | 4.20 | 60.90 | 172.06 | 34.94 | 267.90 |
| extra over for | | | | | | | |
| running outlet | 2 | nr | 0.60 | 8.70 | 21.92 | 4.59 | 35.21 |
| **Total** | | | 14.60 | 211.70 | 498.94 | 106.60 | 817.24 |

| | Qty | Unit | Hours | Hours £ | Mat'ls £ | O & P £ | Total £ |
|---|---|---|---|---|---|---|---|
| **Two-storey semi-detached house size 8 x 7m with gable ends** | | | | | | | |
| Take down existing rainwater goods and make good | | Item | 7.00 | 101.50 | 0.00 | 15.23 | 116.73 |
| 75mm diameter cast iron rainwater pipe with pipe clips plugged to brickwork at 2m maximum centres | 12 | m | 3.60 | 52.20 | 246.12 | 44.75 | 343.07 |
| **Carried forward** | | | 10.60 | 153.70 | 246.12 | 59.97 | 459.79 |

| | Qty | Unit | Hours | Hours £ | Mat'ls £ | O & P £ | Total £ |
|---|---|---|---|---|---|---|---|
| **Brought forward** | | | 10.60 | 153.70 | 246.12 | 59.97 | 459.79 |
| extra over for | | | | | | | |
| offset | 2 | nr | 0.60 | 8.70 | 30.56 | 5.89 | 45.15 |
| shoe | 2 | nr | 0.60 | 8.70 | 28.28 | 5.55 | 42.53 |
| 100mm half round cast iron gutter to softwood fascia | 16 | m | 4.80 | 69.60 | 196.64 | 39.94 | 306.18 |
| extra over for | | | | | | | |
| stop end outlet | 2 | nr | 0.30 | 4.35 | 12.60 | 2.54 | 19.49 |
| **Total** | | | 16.90 | 245.05 | 514.20 | 113.89 | 873.14 |

| | Qty | Unit | Hours | Hours £ | Mat'ls £ | O & P £ | Total £ |
|---|---|---|---|---|---|---|---|
| **Two-storey detached house size 9 x 8m with gable ends** | | | | | | | |
| Take down existing rainwater goods and make good | | Item | 8.00 | 116.00 | 0.00 | 17.40 | 133.40 |
| 75mm diameter cast iron rainwater pipe with pipe clips plugged to brickwork at 2m maximum centres | 12 | m | 3.60 | 52.20 | 246.12 | 44.75 | 343.07 |
| extra over for | | | | | | | |
| offset | 2 | nr | 0.60 | 8.70 | 30.56 | 5.89 | 45.15 |
| shoe | 2 | nr | 0.60 | 8.70 | 28.28 | 5.55 | 42.53 |
| 100mm half round cast iron gutter to softwood fascia | 18 | m | 5.40 | 78.30 | 221.22 | 44.93 | 344.45 |
| extra over for | | | | | | | |
| stop end outlet | 2 | nr | 0.30 | 4.35 | 12.60 | 2.54 | 19.49 |
| **Total** | | | 18.50 | 268.25 | 538.78 | 121.05 | 928.08 |

| | Qty | Unit | Hours | Hours £ | Mat'ls £ | O & P £ | Total £ |
|---|---|---|---|---|---|---|---|
| **Three-storey terraced house size 8 x 8m with gable ends** | | | | | | | |
| Take down existing rainwater goods and make good | | Item | 6.00 | 87.00 | 0.00 | 13.05 | 100.05 |
| 75mm diameter cast iron rainwater pipe with pipe clips plugged to brickwork at 2m maximum centres | 18 | m | 5.40 | 78.30 | 369.18 | 67.12 | 514.60 |
| extra over for | | | | | | | |
| offset | 2 | nr | 0.60 | 8.70 | 30.56 | 5.89 | 45.15 |
| shoe | 2 | nr | 0.60 | 8.70 | 28.28 | 5.55 | 42.53 |
| 100mm half round cast iron gutter to softwood fascia | 16 | m | 1.80 | 26.10 | 196.64 | 33.41 | 256.15 |
| extra over for | | | | | | | |
| running outlet | 2 | nr | 0.60 | 8.70 | 21.92 | 4.59 | 35.21 |
| **Total** | | | 15.00 | 217.50 | 646.58 | 129.61 | 993.69 |

| | Qty | Unit | Hours | Hours £ | Mat'ls £ | O & P £ | Total £ |
|---|---|---|---|---|---|---|---|
| **Three-storey semi-detached house size 9 x 8m with gable ends** | | | | | | | |
| Take down existing rainwater goods and make good | | Item | 7.00 | 101.50 | 0.00 | 15.23 | 116.73 |
| 75mm diameter cast iron rainwater pipe with pipe clips plugged to brickwork at 2m maximum centres | 18 | m | 5.40 | 78.30 | 369.18 | 67.12 | 514.60 |
| **Carried forward** | | | 12.40 | 179.80 | 369.18 | 82.35 | 631.33 |

| | Qty | Unit | Hours | Hours £ | Mat'ls £ | O & P £ | Total £ |
|---|---|---|---|---|---|---|---|
| **Brought forward** | | | 12.40 | 179.80 | 369.18 | 82.35 | 631.33 |
| extra over for | | | | | | | |
| offset | 2 | nr | 0.60 | 8.70 | 30.56 | 5.89 | 45.15 |
| shoe | 2 | nr | 0.60 | 8.70 | 28.28 | 5.55 | 42.53 |
| 100mm half round cast iron gutter to softwood fascia | 18 | m | 5.40 | 78.30 | 221.22 | 44.93 | 344.45 |
| extra over for | | | | | | | |
| stop end outlet | 2 | nr | 0.30 | 4.35 | 12.60 | 2.54 | 19.49 |
| **Total** | | | 19.30 | 279.85 | 661.84 | 141.25 | 1,082.94 |

| | Qty | Unit | Hours | Hours £ | Mat'ls £ | O & P £ | Total £ |
|---|---|---|---|---|---|---|---|
| **Three-storey detached house size 10 x 9m with gable ends** | | | | | | | |
| Take down existing rainwater goods and make good | | Item | 9.00 | 130.50 | 0.00 | 19.58 | 150.08 |
| 75mm diameter cast iron rainwater pipe with pipe clips plugged to brickwork at 2m maximum centres | 18 | m | 5.40 | 78.30 | 369.18 | 67.12 | 514.60 |
| extra over for | | | | | | | |
| offset | 2 | nr | 0.60 | 8.70 | 30.56 | 5.89 | 45.15 |
| shoe | 2 | nr | 0.60 | 8.70 | 28.28 | 5.55 | 42.53 |
| 100mm half round cast iron gutter to softwood fascia | 20 | m | 6.00 | 87.00 | 245.80 | 49.92 | 382.72 |
| extra over for | | | | | | | |
| stop end outlet | 2 | nr | 0.30 | 4.35 | 12.60 | 2.54 | 19.49 |
| **Total** | | | 21.90 | 317.55 | 686.42 | 150.60 | 1,154.57 |

| | Qty | Unit | Hours | Hours £ | Mat'ls £ | O & P £ | Total £ |
|---|---|---|---|---|---|---|---|
| **One-storey terraced house size 7 x 6m with hipped ends** | | | | | | | |
| Take down existing rainwater goods and make good | | Item | 4.00 | 58.00 | 0.00 | 8.70 | 66.70 |
| 75mm diameter cast iron rainwater pipe with pipe clips plugged to brickwork at 2m maximum centres | 6 | m | 1.80 | 26.10 | 123.06 | 22.37 | 171.53 |
| extra over for | | | | | | | |
| offset | 2 | nr | 0.60 | 8.70 | 30.56 | 5.89 | 45.15 |
| shoe | 2 | nr | 0.60 | 8.70 | 28.28 | 5.55 | 42.53 |
| 100mm half round cast iron gutter to softwood fascia | 14 | m | 4.20 | 60.90 | 172.06 | 34.94 | 267.90 |
| extra over for | | | | | | | |
| running outlet | 2 | nr | 0.60 | 8.70 | 21.92 | 4.59 | 35.21 |
| **Total** | | | 11.80 | 171.10 | 375.88 | 82.05 | 629.03 |
| **One-storey semi-detached house size 8 x 7m with hipped ends** | | | | | | | |
| Take down existing rainwater goods and make good | | Item | 6.00 | 87.00 | 0.00 | 13.05 | 100.05 |
| 75mm diameter cast iron rainwater pipe with pipe clips plugged to brickwork at 2m maximum centres | 6 | m | 1.80 | 26.10 | 85.80 | 16.79 | 128.69 |
| extra over for | | | | | | | |
| offset | 2 | nr | 0.60 | 8.70 | 30.56 | 5.89 | 45.15 |
| shoe | 2 | nr | 0.60 | 8.70 | 28.28 | 5.55 | 42.53 |
| **Carried forward** | | | 9.00 | 130.50 | 144.64 | 41.27 | 316.41 |

| | Qty | Unit | Hours | Hours £ | Mat'ls £ | O & P £ | Total £ |
|---|---|---|---|---|---|---|---|
| **Brought forward** | | | 9.00 | 130.50 | 144.64 | 41.27 | 316.41 |
| 100mm half round cast iron gutter to softwood fascia | 23 | m | 6.90 | 100.05 | 282.67 | 57.41 | 440.13 |
| extra over for | | | | | | | |
| angle | 2 | nr | 0.60 | 8.70 | 21.92 | 4.59 | 35.21 |
| running outlet | 2 | nr | 0.60 | 8.70 | 12.90 | 3.24 | 24.84 |
| **Total** | | | 17.10 | 247.95 | 462.13 | 106.51 | 816.59 |

| | Qty | Unit | Hours | Hours £ | Mat'ls £ | O & P £ | Total £ |
|---|---|---|---|---|---|---|---|
| **One-storey detached house size 9 x 7m with hipped ends** | | | | | | | |
| Take down existing rainwater goods and make good | | Item | 8.00 | 116.00 | 0.00 | 17.40 | 133.40 |
| 75mm diameter cast iron rainwater pipe with pipe clips plugged to brickwork at 2m maximum centres | 6 | m | 1.80 | 26.10 | 123.06 | 22.37 | 171.53 |
| extra over for | | | | | | | |
| offset | 2 | nr | 0.60 | 8.70 | 30.56 | 5.89 | 45.15 |
| shoe | 2 | nr | 0.60 | 8.70 | 28.28 | 5.55 | 42.53 |
| 100mm half round cast iron gutter to softwood fascia | 32 | m | 9.60 | 139.20 | 393.28 | 79.87 | 612.35 |
| extra over for | | | | | | | |
| angle | 4 | nr | 0.60 | 8.70 | 21.92 | 4.59 | 35.21 |
| running outlet | 2 | nr | 1.20 | 17.40 | 25.80 | 6.48 | 49.68 |
| **Total** | | | 22.40 | 324.80 | 622.90 | 142.16 | 1,089.86 |

| | Qty | Unit | Hours | Hours £ | Mat'ls £ | O & P £ | Total £ |
|---|---|---|---|---|---|---|---|
| **Two-storey terraced house size 7 x 7m with hipped ends** | | | | | | | |
| Take down existing rainwater goods and make good | | Item | 6.00 | 87.00 | 0.00 | 13.05 | 100.05 |
| 75mm diameter cast iron rainwater pipe with pipe clips plugged to brickwork at 2m maximum centres | 12 | m | 3.60 | 52.20 | 246.12 | 44.75 | 343.07 |
| extra over for | | | | | | | |
| offset | 2 | nr | 0.60 | 8.70 | 30.56 | 5.89 | 45.15 |
| shoe | 2 | nr | 0.60 | 8.70 | 28.28 | 5.55 | 42.53 |
| 100mm half round cast iron gutter to softwood fascia | 14 | m | 4.20 | 60.90 | 172.06 | 34.94 | 267.90 |
| extra over for | | | | | | | |
| running outlet | 2 | nr | 0.60 | 8.70 | 21.92 | 4.59 | 35.21 |
| **Total** | | | 15.60 | 226.20 | 498.94 | 108.77 | 833.91 |

| | Qty | Unit | Hours | Hours £ | Mat'ls £ | O & P £ | Total £ |
|---|---|---|---|---|---|---|---|
| **Two-storey semi-detached house size 8 x 7m with hipped ends** | | | | | | | |
| Take down existing rainwater goods and make good | | Item | 8.00 | 116.00 | 0.00 | 17.40 | 133.40 |
| 75mm diameter cast iron rainwater pipe with pipe clips plugged to brickwork at 2m maximum centres | 12 | m | 3.60 | 52.20 | 246.12 | 44.75 | 343.07 |
| **Carried forward** | | | 11.60 | 168.20 | 246.12 | 62.15 | 476.47 |

| | Qty | Unit | Hours | Hours £ | Mat'ls £ | O & P £ | Total £ |
|---|---|---|---|---|---|---|---|
| **Brought forward** | | | 11.60 | 168.20 | 246.12 | 62.15 | 476.47 |
| extra over for | | | | | | | |
| offset | 2 | nr | 0.60 | 8.70 | 30.56 | 5.89 | 45.15 |
| shoe | 2 | nr | 0.60 | 8.70 | 28.28 | 5.55 | 42.53 |
| 100mm half round cast iron gutter to softwood fascia | 23 | m | 6.90 | 100.05 | 282.67 | 57.41 | 440.13 |
| extra over for | | | | | | | |
| angle | 2 | nr | 0.60 | 8.70 | 21.92 | 4.59 | 35.21 |
| running outlet | 2 | nr | 0.60 | 8.70 | 12.90 | 3.24 | 24.84 |
| **Total** | | | 20.90 | 303.05 | 622.45 | 138.83 | 1,064.33 |

| | Qty | Unit | Hours | Hours £ | Mat'ls £ | O & P £ | Total £ |
|---|---|---|---|---|---|---|---|
| **Two-storey detached house size 9 x 8m with hipped ends** | | | | | | | |
| Take down existing rainwater goods and make good | | Item | 9.00 | 130.50 | 0.00 | 19.58 | 150.08 |
| 75mm diameter cast iron rainwater pipe as before | 12 | m | 3.60 | 52.20 | 246.12 | 44.75 | 343.07 |
| extra over for | | | | | | | |
| offset | 2 | nr | 0.60 | 8.70 | 30.56 | 5.89 | 45.15 |
| shoe | 2 | nr | 0.60 | 8.70 | 28.28 | 5.55 | 42.53 |
| 100mm half round cast iron gutter to softwood fascia | 34 | m | 10.20 | 147.90 | 417.86 | 84.86 | 650.62 |
| extra over for | | | | | | | |
| angle | 4 | nr | 0.60 | 8.70 | 21.92 | 4.59 | 35.21 |
| running outlet | 2 | nr | 1.20 | 17.40 | 25.80 | 6.48 | 49.68 |
| **Total** | | | 25.80 | 374.10 | 770.54 | 171.70 | 1,316.34 |

| | Qty | Unit | Hours | Hours £ | Mat'ls £ | O & P £ | Total £ |
|---|---|---|---|---|---|---|---|
| **Three-storey terraced house size 8 x 8m with hipped ends** | | | | | | | |
| Take down existing rainwater goods and make good | | Item | 7.00 | 101.50 | 0.00 | 15.23 | 116.73 |
| 75mm diameter cast iron rainwater pipe with pipe clips plugged to brickwork at 2m maximum centres | 18 | m | 5.40 | 78.30 | 369.18 | 67.12 | 514.60 |
| extra over for | | | | | | | |
| offset | 2 | nr | 0.60 | 8.70 | 30.56 | 5.89 | 45.15 |
| shoe | 2 | nr | 0.60 | 8.70 | 28.28 | 5.55 | 42.53 |
| 100mm half round cast iron gutter to softwood fascia | 16 | m | 4.80 | 69.60 | 196.64 | 39.94 | 306.18 |
| extra over for | | | | | | | |
| running outlet | 2 | nr | 0.60 | 8.70 | 21.92 | 4.59 | 35.21 |
| **Total** | | | 19.00 | 275.50 | 646.58 | 138.31 | 1,060.39 |

| | Qty | Unit | Hours | Hours £ | Mat'ls £ | O & P £ | Total £ |
|---|---|---|---|---|---|---|---|
| **Three-storey semi-detached house size 9 x 8m with hipped ends** | | | | | | | |
| Take down existing rainwater goods and make good | | Item | 9.00 | 130.50 | 0.00 | 19.58 | 150.08 |
| 75mm diameter cast iron rainwater pipe with pipe clips plugged to brickwork at 2m maximum centres | 18 | m | 5.40 | 78.30 | 369.18 | 67.12 | 514.60 |
| **Carried forward** | | | 14.40 | 208.80 | 369.18 | 86.70 | 664.68 |

| | Qty | Unit | Hours | Hours £ | Mat'ls £ | O & P £ | Total £ |
|---|---|---|---|---|---|---|---|
| **Brought forward** | | | 14.40 | 208.80 | 369.18 | 86.70 | 664.68 |
| extra over for | | | | | | | |
| offset | 2 | nr | 0.60 | 8.70 | 30.56 | 5.89 | 45.15 |
| shoe | 2 | nr | 0.60 | 8.70 | 28.28 | 5.55 | 42.53 |
| 100mm half round cast iron gutter to softwood fascia | 26 | m | 7.80 | 113.10 | 319.56 | 64.90 | 497.56 |
| extra over for | | | | | | | |
| angle | 2 | nr | 0.60 | 8.70 | 21.92 | 4.59 | 35.21 |
| running outlet | 2 | nr | 0.60 | 8.70 | 12.90 | 3.24 | 24.84 |
| **Total** | | | 24.60 | 356.70 | 782.40 | 170.87 | 1,309.97 |

| | Qty | Unit | Hours | Hours £ | Mat'ls £ | O & P £ | Total £ |
|---|---|---|---|---|---|---|---|
| **Three-storey detached house size 10 x 9m with hipped ends** | | | | | | | |
| Take down existing rainwater goods and make good | | Item | 10.00 | 145.00 | 0.00 | 21.75 | 166.75 |
| 75mm diameter cast iron rainwater pipe as before | 18 | m | 5.40 | 78.30 | 369.18 | 67.12 | 514.60 |
| extra over for | | | | | | | |
| offset | 2 | nr | 0.60 | 8.70 | 30.56 | 5.89 | 45.15 |
| shoe | 2 | nr | 0.60 | 8.70 | 28.28 | 5.55 | 42.53 |
| 100mm half round cast iron gutter to softwood fascia | 38 | m | 11.40 | 165.30 | 467.02 | 94.85 | 727.17 |
| extra over for | | | | | | | |
| angle | 4 | nr | 0.60 | 8.70 | 21.92 | 4.59 | 35.21 |
| running outlet | 2 | nr | 1.20 | 17.40 | 25.80 | 6.48 | 49.68 |
| **Total** | | | 29.80 | 432.10 | 942.76 | 206.23 | 1,581.09 |

| | Qty | Unit | Hours | Hours £ | Mat'ls £ | O & P £ | Total £ |
|---|---|---|---|---|---|---|---|
| **ALUMINIUM RAINWATER GOODS** | | | | | | | |
| **One-storey terraced house size 7 x 6m with gable ends** | | | | | | | |
| Take down existing rainwater goods and make good | | Item | 4.00 | 58.00 | 0.00 | 8.70 | 66.70 |
| 76mm diameter aluminium rainwater pipe with pipe clips plugged to brickwork at 2m maximum centres | 6 | m | 1.80 | 26.10 | 113.34 | 20.92 | 160.36 |
| extra over for | | | | | | | |
| offset | 2 | nr | 0.60 | 8.70 | 53.08 | 9.27 | 71.05 |
| shoe | 2 | nr | 0.60 | 8.70 | 24.32 | 4.95 | 37.97 |
| 100mm half round aluminium gutter to softwood fascia | 14 | m | 3.20 | 46.40 | 153.44 | 29.98 | 229.82 |
| extra over for | | | | | | | |
| running outlet | 2 | nr | 0.60 | 8.70 | 21.92 | 4.59 | 35.21 |
| **Total** | | | 10.80 | 156.60 | 366.10 | 78.41 | 601.11 |
| **One-storey semi-detached house size 8 x 7m with gable ends** | | | | | | | |
| Take down existing rainwater goods and make good | | Item | 6.00 | 87.00 | 0.00 | 13.05 | 100.05 |
| 76mm diameter aluminium rainwater pipe with pipe clips plugged to brickwork at 2m maximum centres | 6 | m | 1.80 | 26.10 | 113.34 | 20.92 | 160.36 |
| **Carried forward** | | | 7.80 | 113.10 | 113.34 | 33.97 | 260.41 |

| | Qty | Unit | Hours | Hours £ | Mat'ls £ | O & P £ | Total £ |
|---|---|---|---|---|---|---|---|
| **Brought forward** | | | 7.80 | 113.10 | 113.34 | 33.97 | 260.41 |
| extra over for | | | | | | | |
| offset | 2 | nr | 0.60 | 8.70 | 53.08 | 9.27 | 71.05 |
| shoe | 2 | nr | 0.60 | 8.70 | 24.32 | 4.95 | 37.97 |
| 100mm half round aluminium gutter to softwood fascia | 16 | m | 4.80 | 69.60 | 175.36 | 36.74 | 281.70 |
| extra over for | | | | | | | |
| stop end outlet | 2 | nr | 0.30 | 4.35 | 12.60 | 2.54 | 19.49 |
| **Total** | | | 14.10 | 204.45 | 378.70 | 87.48 | 670.63 |

| | Qty | Unit | Hours | Hours £ | Mat'ls £ | O & P £ | Total £ |
|---|---|---|---|---|---|---|---|
| **One-storey detached house size 9 x 7m with gable ends** | | | | | | | |
| Take down existing rainwater goods and make good | | Item | 7.00 | 101.50 | 0.00 | 15.23 | 116.73 |
| 76mm diameter aluminium rainwater pipe with pipe clips plugged to brickwork at 2m maximum centres | 6 | m | 1.80 | 26.10 | 113.34 | 20.92 | 160.36 |
| extra over for | | | | | | | |
| offset | 2 | nr | 0.60 | 8.70 | 53.08 | 9.27 | 71.05 |
| shoe | 2 | nr | 0.60 | 8.70 | 24.32 | 4.95 | 37.97 |
| 100mm half round aluminium gutter to softwood fascia | 32 | m | 5.40 | 78.30 | 197.28 | 41.34 | 316.92 |
| extra over for | | | | | | | |
| running outlet | 2 | nr | 0.30 | 4.35 | 12.60 | 2.54 | 19.49 |
| **Total** | | | 15.70 | 227.65 | 400.62 | 94.24 | 722.51 |

| | Qty | Unit | Hours | Hours £ | Mat'ls £ | O & P £ | Total £ |
|---|---|---|---|---|---|---|---|
| **Two-storey terraced house size 7 x 7m with gable ends** | | | | | | | |
| Take down existing rainwater goods and make good | | Item | 5.00 | 72.50 | 0.00 | 10.88 | 83.38 |
| 76mm diameter aluminium rainwater pipe with pipe clips plugged to brickwork at 2m maximum centres | 12 | m | 3.60 | 52.20 | 226.68 | 41.83 | 320.71 |
| extra over for | | | | | | | |
| offset | 2 | nr | 0.60 | 8.70 | 53.08 | 9.27 | 71.05 |
| shoe | 2 | nr | 0.60 | 8.70 | 24.32 | 4.95 | 37.97 |
| 100mm half round aluminium gutter to softwood fascia | 14 | m | 3.20 | 46.40 | 153.34 | 29.96 | 229.70 |
| extra over for | | | | | | | |
| running outlet | 2 | nr | 0.60 | 8.70 | 21.92 | 4.59 | 35.21 |
| **Total** | | | 13.60 | 197.20 | 479.34 | 101.48 | 778.02 |

| | Qty | Unit | Hours | Hours £ | Mat'ls £ | O & P £ | Total £ |
|---|---|---|---|---|---|---|---|
| **Two-storey semi-detached house size 8 x 7m with gable ends** | | | | | | | |
| Take down existing rainwater goods and make good | | Item | 7.00 | 101.50 | 0.00 | 15.23 | 116.73 |
| 76mm diameter aluminium rainwater pipe with pipe clips plugged to brickwork at 2m maximum centres | 12 | m | 3.60 | 52.20 | 226.28 | 41.77 | 320.25 |
| **Carried forward** | | | 10.60 | 153.70 | 226.28 | 57.00 | 436.98 |

| | Qty | Unit | Hours | Hours £ | Mat'ls £ | O & P £ | Total £ |
|---|---|---|---|---|---|---|---|
| **Brought forward** | | | 10.60 | 153.70 | 226.28 | 57.00 | 436.98 |
| extra over for | | | | | | | |
| offset | 2 | nr | 0.60 | 8.70 | 53.08 | 9.27 | 71.05 |
| shoe | 2 | nr | 0.60 | 8.70 | 24.32 | 4.95 | 37.97 |
| 100mm half round aluminium gutter to softwood fascia | 16 | m | 4.80 | 69.60 | 175.36 | 36.74 | 281.70 |
| extra over for | | | | | | | |
| stop end outlet | 2 | nr | 0.30 | 4.35 | 12.60 | 2.54 | 19.49 |
| **Total** | | | 16.90 | 245.05 | 491.64 | 110.50 | 847.19 |

| | Qty | Unit | Hours | Hours £ | Mat'ls £ | O & P £ | Total £ |
|---|---|---|---|---|---|---|---|
| **Two-storey detached house size 9 x 8m with gable ends** | | | | | | | |
| Take down existing rainwater goods and make good | | Item | 8.00 | 116.00 | 0.00 | 17.40 | 133.40 |
| 76mm diameter aluminium rainwater pipe with pipe clips plugged to brickwork at 2m maximum centres | 12 | m | 3.60 | 52.20 | 226.68 | 41.83 | 320.71 |
| extra over for | | | | | | | |
| offset | 2 | nr | 0.60 | 8.70 | 53.08 | 9.27 | 71.05 |
| shoe | 2 | nr | 0.60 | 8.70 | 24.32 | 4.95 | 37.97 |
| 100mm half round aluminium gutter to softwood fascia | 18 | m | 5.40 | 78.30 | 197.28 | 41.34 | 316.92 |
| extra over for | | | | | | | |
| stop end outlet | 2 | nr | 0.30 | 4.35 | 12.60 | 2.54 | 19.49 |
| **Total** | | | 18.50 | 268.25 | 513.96 | 117.33 | 899.54 |

| | Qty | Unit | Hours | Hours £ | Mat'ls £ | O & P £ | Total £ |
|---|---|---|---|---|---|---|---|
| **Three-storey terraced house size 8 x 8m with gable ends** | | | | | | | |
| Take down existing rainwater goods and make good | | Item | 6.00 | 87.00 | 0.00 | 13.05 | 100.05 |
| 76mm diameter aluminium rainwater pipe with pipe clips plugged to brickwork at 2m maximum centres | 18 | m | 5.40 | 78.30 | 340.02 | 62.75 | 481.07 |
| extra over for | | | | | | | |
| offset | 2 | nr | 0.60 | 8.70 | 53.08/ | 1.31 | 10.01 |
| shoe | 2 | nr | 0.60 | 8.70 | 24.32 | 4.95 | 37.97 |
| 100mm half round aluminium gutter to softwood fascia | 16 | m | 4.80 | 69.60 | 175.36 | 36.74 | 281.70 |
| extra over for | | | | | | | |
| running outlet | 2 | nr | 0.60 | 8.70 | 21.92 | 4.59 | 35.21 |
| **Total** | | | 18.00 | 261.00 | 561.62 | 123.39 | 946.01 |

| | Qty | Unit | Hours | Hours £ | Mat'ls £ | O & P £ | Total £ |
|---|---|---|---|---|---|---|---|
| **Three-storey semi-detached house size 9 x 8m with gable ends** | | | | | | | |
| Take down existing rainwater goods and make good | | Item | 8.00 | 116.00 | 0.00 | 17.40 | 133.40 |
| 75mm diameter aluminium rainwater pipe with pipe clips plugged to brickwork at 2m maximum centres | 18 | m | 5.40 | 78.30 | 340.02 | 62.75 | 481.07 |
| **Carried forward** | | | 13.40 | 194.30 | 340.02 | 80.15 | 614.47 |

| | Qty | Unit | Hours | Hours £ | Mat'ls £ | O & P £ | Total £ |
|---|---|---|---|---|---|---|---|
| **Brought forward** | | | 13.40 | 194.30 | 340.02 | 80.15 | 614.47 |
| extra over for | | | | | | | |
| offset | 2 | nr | 0.60 | 8.70 | 53.08 | 9.27 | 71.05 |
| shoe | 2 | nr | 0.60 | 8.70 | 24.32 | 4.95 | 37.97 |
| 100mm half round aluminium gutter to softwood fascia | 18 | m | 5.40 | 78.30 | 197.28 | 41.34 | 316.92 |
| extra over for | | | | | | | |
| stop end outlet | 2 | nr | 0.30 | 4.35 | 12.60 | 2.54 | 19.49 |
| **Total** | | | 20.30 | 294.35 | 627.30 | 138.25 | 1,059.90 |

| | Qty | Unit | Hours | Hours £ | Mat'ls £ | O & P £ | Total £ |
|---|---|---|---|---|---|---|---|
| **Three-storey detached house size 10 x 9m with gable ends** | | | | | | | |
| Take down existing rainwater goods and make good | | Item | 9.00 | 130.50 | 0.00 | 19.58 | 150.08 |
| 75mm diameter aluminium rainwater pipe with pipe clips plugged to brickwork at 2m maximum centres | 18 | m | 5.40 | 78.30 | 340.02 | 62.75 | 481.07 |
| extra over for | | | | | | | |
| offset | 2 | nr | 0.60 | 8.70 | 53.08 | 9.27 | 71.05 |
| shoe | 2 | nr | 0.60 | 8.70 | 24.32 | 4.95 | 37.97 |
| 100mm half round aluminium gutter to softwood fascia | 20 | m | 6.00 | 87.00 | 219.20 | 45.93 | 352.13 |
| extra over for | | | | | | | |
| stop end outlet | 2 | nr | 0.30 | 4.35 | 12.60 | 2.54 | 19.49 |
| **Total** | | | 21.90 | 317.55 | 649.22 | 145.02 | 1,111.79 |

| | Qty | Unit | Hours | Hours £ | Mat'ls £ | O & P £ | Total £ |
|---|---|---|---|---|---|---|---|
| **One-storey terraced house size 7 x 6m with hipped ends** | | | | | | | |
| Take down existing rainwater goods and make good | | Item | 5.00 | 72.50 | 0.00 | 10.88 | 83.38 |
| 75mm diameter aluminium rainwater pipe with pipe clips plugged to brickwork at 2m maximum centres | 6 | m | 1.80 | 26.10 | 113.34 | 20.92 | 160.36 |
| extra over for | | | | | | | |
| offset | 2 | nr | 0.60 | 8.70 | 53.08 | 9.27 | 71.05 |
| shoe | 2 | nr | 0.60 | 8.70 | 24.32 | 4.95 | 37.97 |
| 100mm half round aluminium gutter to softwood fascia | 14 | m | 3.20 | 46.40 | 152.44 | 29.83 | 228.67 |
| extra over for | | | | | | | |
| running outlet | 2 | nr | 0.60 | 8.70 | 21.92 | 4.59 | 35.21 |
| **Total** | | | 11.80 | 171.10 | 365.10 | 80.43 | 616.63 |
| **One-storey semi-detached house size 8 x 7m with hipped ends** | | | | | | | |
| Take down existing rainwater goods and make good | | Item | 6.00 | 87.00 | 0.00 | 13.05 | 100.05 |
| 75mm diameter aluminium rainwater pipe with pipe clips plugged to brickwork at 2m maximum centres | 6 | m | 1.80 | 26.10 | 113.34 | 20.92 | 160.36 |
| extra over for | | | | | | | |
| offset | 2 | nr | 0.60 | 8.70 | 53.08 | 9.27 | 71.05 |
| shoe | 2 | nr | 0.60 | 8.70 | 24.32 | 4.95 | 37.97 |
| **Carried forward** | | | 9.00 | 130.50 | 190.74 | 48.19 | 369.43 |

| | Qty | Unit | Hours | Hours £ | Mat'ls £ | O & P £ | Total £ |
|---|---|---|---|---|---|---|---|
| **Brought forward** | | | 9.00 | 130.50 | 190.74 | 48.19 | 369.43 |
| 100mm half round aluminium gutter to softwood fascia | 23 | m | 6.90 | 100.05 | 282.67 | 57.41 | 440.13 |
| extra over for | | | | | | | |
| angle | 2 | nr | 0.60 | 8.70 | 21.92 | 4.59 | 35.21 |
| running outlet | 2 | nr | 0.60 | 8.70 | 12.90 | 3.24 | 24.84 |
| **Total** | | | 17.10 | 247.95 | 508.23 | 113.43 | 869.61 |

| | Qty | Unit | Hours | Hours £ | Mat'ls £ | O & P £ | Total £ |
|---|---|---|---|---|---|---|---|
| **One-storey detached house size 9 x 7m with hipped ends** | | | | | | | |
| Take down existing rainwater goods and make good | | Item | 8.00 | 116.00 | 0.00 | 17.40 | 133.40 |
| 75mm diameter aluminium rainwater pipe with pipe clips plugged to brickwork at 2m maximum centres | 6 | m | 1.80 | 26.10 | 113.34 | 20.92 | 160.36 |
| extra over for | | | | | | | |
| offset | 2 | nr | 0.60 | 8.70 | 53.08 | 9.27 | 71.05 |
| shoe | 2 | nr | 0.60 | 8.70 | 24.32 | 4.95 | 37.97 |
| 100mm half round aluminium gutter to softwood fascia | 32 | m | 9.60 | 139.20 | 393.28 | 79.87 | 612.35 |
| extra over for | | | | | | | |
| angle | 4 | nr | 0.60 | 8.70 | 21.92 | 4.59 | 35.21 |
| running outlet | 2 | nr | 1.20 | 17.40 | 25.80 | 6.48 | 49.68 |
| **Total** | | | 22.40 | 324.80 | 631.74 | 143.48 | 1,100.02 |

| | Qty | Unit | Hours | Hours £ | Mat'ls £ | O & P £ | Total £ |
|---|---|---|---|---|---|---|---|
| **Two-storey terraced house size 7 x 7m with hipped ends** | | | | | | | |
| Take down existing rainwater goods and make good | | Item | 6.00 | 87.00 | 0.00 | 13.05 | 100.05 |
| 75mm diameter aluminium rainwater pipe with pipe clips plugged to brickwork at 2m maximum centres | 12 | m | 3.60 | 52.20 | 226.68 | 41.83 | 320.71 |
| extra over for | | | | | | | |
| offset | 2 | nr | 0.60 | 8.70 | 53.08 | 9.27 | 71.05 |
| shoe | 2 | nr | 0.60 | 8.70 | 24.32 | 4.95 | 37.97 |
| 100mm half round aluminium gutter to softwood fascia | 14 | m | 3.20 | 46.40 | 152.44 | 29.83 | 228.67 |
| extra over for | | | | | | | |
| running outlet | 2 | nr | 0.60 | 8.70 | 21.92 | 4.59 | 35.21 |
| **Total** | | | 14.60 | 211.70 | 478.44 | 103.52 | 793.66 |

| | Qty | Unit | Hours | Hours £ | Mat'ls £ | O & P £ | Total £ |
|---|---|---|---|---|---|---|---|
| **Two-storey semi-detached house size 8 x 7m with hipped ends** | | | | | | | |
| Take down existing rainwater goods and make good | | Item | 8.00 | 116.00 | 0.00 | 17.40 | 133.40 |
| 75mm diameter aluminium rainwater pipe with pipe clips plugged to brickwork at 2m maximum centres | 12 | m | 3.60 | 52.20 | 226.68 | 41.83 | 320.71 |
| **Carried forward** | | | 11.60 | 168.20 | 226.68 | 59.23 | 454.11 |

| | Qty | Unit | Hours | Hours £ | Mat'ls £ | O & P £ | Total £ |
|---|---|---|---|---|---|---|---|
| **Brought forward** | | | 11.60 | 168.20 | 226.68 | 59.23 | 454.11 |
| extra over for | | | | | | | |
| offset | 2 | nr | 0.60 | 8.70 | 53.08 | 9.27 | 71.05 |
| shoe | 2 | nr | 0.60 | 8.70 | 24.32 | 4.95 | 37.97 |
| 100mm half round aluminium gutter to softwood fascia | 23 | m | 6.90 | 100.05 | 282.67 | 57.41 | 440.13 |
| extra over for | | | | | | | |
| angle | 2 | nr | 0.60 | 8.70 | 21.92 | 4.59 | 35.21 |
| running outlet | 2 | nr | 0.60 | 8.70 | 12.90 | 3.24 | 24.84 |
| **Total** | | | 20.90 | 303.05 | 621.57 | 138.69 | 1,063.31 |

| | Qty | Unit | Hours | Hours £ | Mat'ls £ | O & P £ | Total £ |
|---|---|---|---|---|---|---|---|
| **Two-storey detached house size 9 x 8m with hipped ends** | | | | | | | |
| Take down existing rainwater goods and make good | | Item | 9.00 | 130.50 | 0.00 | 19.58 | 150.08 |
| 75mm diameter aluminium rainwater pipe as before | 12 | m | 3.60 | 52.20 | 226.68 | 41.83 | 320.71 |
| extra over for | | | | | | | |
| offset | 2 | nr | 0.60 | 8.70 | 53.08 | 9.27 | 71.05 |
| shoe | 2 | nr | 0.60 | 8.70 | 24.32 | 4.95 | 37.97 |
| 100mm half round aluminium gutter to softwood fascia | 34 | m | 10.20 | 147.90 | 417.86 | 84.86 | 650.62 |
| extra over for | | | | | | | |
| angle | 4 | nr | 0.60 | 8.70 | 21.92 | 4.59 | 35.21 |
| running outlet | 2 | nr | 1.20 | 17.40 | 25.80 | 6.48 | 49.68 |
| **Total** | | | 25.80 | 374.10 | 769.66 | 171.56 | 1,315.32 |

| | Qty | Unit | Hours | Hours £ | Mat'ls £ | O & P £ | Total £ |
|---|---|---|---|---|---|---|---|
| **Three-storey terraced house size 8 x 8m with hipped ends** | | | | | | | |
| Take down existing rainwater goods and make good | | Item | 7.00 | 101.50 | 0.00 | 15.23 | 116.73 |
| 75mm diameter aluminium rainwater pipe with pipe clips plugged to brickwork at 2m maximum centres | 18 | m | 5.40 | 78.30 | 340.02 | 62.75 | 481.07 |
| extra over for | | | | | | | |
| offset | 2 | nr | 0.60 | 8.70 | 53.08 | 9.27 | 71.05 |
| shoe | 2 | nr | 0.60 | 8.70 | 24.32 | 4.95 | 37.97 |
| 100mm half round aluminium gutter to softwood fascia | 16 | m | 4.80 | 69.60 | 196.44 | 39.91 | 305.95 |
| extra over for | | | | | | | |
| running outlet | 2 | nr | 0.60 | 8.70 | 21.92 | 4.59 | 35.21 |
| **Total** | | | 19.00 | 275.50 | 635.78 | 136.69 | 1,047.97 |

| | Qty | Unit | Hours | Hours £ | Mat'ls £ | O & P £ | Total £ |
|---|---|---|---|---|---|---|---|
| **Three-storey semi-detached house size 9 x 8m with hipped ends** | | | | | | | |
| Take down existing rainwater goods and make good | | Item | 9.00 | 130.50 | 0.00 | 19.58 | 150.08 |
| 68mm diameter PVC-U rainwater pipe with pipe clips plugged to brickwork at 2m maximum centres | 18 | m | 5.40 | 78.30 | 340.02 | 62.75 | 481.07 |
| **Carried forward** | | | 14.40 | 208.80 | 340.02 | 82.32 | 631.14 |

| | Qty | Unit | Hours | Hours £ | Mat'ls £ | O & P £ | Total £ |
|---|---|---|---|---|---|---|---|
| **Brought forward** | | | 14.40 | 208.80 | 340.02 | 82.32 | 631.14 |
| extra over for | | | | | | | |
| offset | 2 | nr | 0.60 | 8.70 | 53.08 | 9.27 | 71.05 |
| shoe | 2 | nr | 0.60 | 8.70 | 24.32 | 4.95 | 37.97 |
| 100mm half round aluminium gutter to softwood fascia | 26 | m | 7.80 | 113.10 | 319.54 | 64.90 | 497.54 |
| extra over for | | | | | | | |
| angle | 2 | nr | 0.60 | 8.70 | 21.92 | 4.59 | 35.21 |
| running outlet | 2 | nr | 0.60 | 8.70 | 12.90 | 3.24 | 24.84 |
| **Total** | | | 24.60 | 356.70 | 771.78 | 169.27 | 1,297.75 |

| | Qty | Unit | Hours | Hours £ | Mat'ls £ | O & P £ | Total £ |
|---|---|---|---|---|---|---|---|
| **Three-storey detached house size 10 x 9m with hipped ends** | | | | | | | |
| Take down existing rainwater goods and make good | | Item | 10.00 | 145.00 | 0.00 | 21.75 | 166.75 |
| 75mm diameter cast iron rainwater pipe as before | 18 | m | 5.40 | 78.30 | 340.02 | 62.75 | 481.07 |
| extra over for | | | | | | | |
| offset | 2 | nr | 0.60 | 8.70 | 53.08 | 9.27 | 71.05 |
| shoe | 2 | nr | 0.60 | 8.70 | 24.32 | 4.95 | 37.97 |
| 100mm half round cast iron gutter to softwood fascia | 38 | m | 11.40 | 165.30 | 467.02 | 94.85 | 727.17 |
| extra over for | | | | | | | |
| angle | 4 | nr | 0.60 | 8.70 | 21.92 | 4.59 | 35.21 |
| running outlet | 2 | nr | 1.20 | 17.40 | 25.80 | 6.48 | 49.68 |
| **Total** | | | 29.80 | 432.10 | 932.16 | 204.64 | 1,568.90 |

## SUMMARY OF RAINWATER GOODS PROJECT COSTS

| | Hours | Hours £ | Mat'ls £ | O & P £ | Total £ |
|---|---|---|---|---|---|
| **PVC-U RAINWATER GOODS** | | | | | |
| **One-storey gable end** | | | | | |
| Terraced | 9.82 | 142.39 | 119.69 | 39.31 | 301.39 |
| Semi-detached | 11.86 | 171.97 | 124.76 | 44.51 | 341.24 |
| Detached | 13.26 | 192.27 | 133.72 | 48.90 | 374.89 |
| **Two-storey gable end** | | | | | |
| Terraced | 12.32 | 178.64 | 158.62 | 50.59 | 387.85 |
| Semi-detached | 14.86 | 215.47 | 167.58 | 57.46 | 440.51 |
| Detached | 15.86 | 229.97 | 167.59 | 59.63 | 457.19 |
| **Three-storey gable end** | | | | | |
| Terraced | 15.22 | 220.69 | 196.18 | 62.53 | 479.40 |
| Semi-detached | 16.36 | 237.22 | 201.24 | 65.77 | 504.23 |
| Detached | 18.76 | 272.02 | 210.20 | 72.33 | 554.55 |
| **One-storey hipped end** | | | | | |
| Terraced | 9.82 | 142.39 | 119.90 | 39.34 | 301.63 |
| Semi-detached | 14.14 | 205.03 | 153.46 | 53.77 | 412.26 |
| Detached | 18.46 | 267.67 | 218.44 | 72.92 | 559.03 |
| **Two-storey hipped end** | | | | | |
| Terraced | 13.32 | 193.14 | 153.56 | 52.01 | 398.71 |
| Semi-detached | 17.64 | 255.78 | 203.78 | 68.93 | 528.49 |
| Detached | 21.36 | 309.72 | 260.96 | 85.60 | 656.28 |
| **Three-storey hipped end** | | | | | |
| Terraced | 16.22 | 235.19 | 196.18 | 64.71 | 496.08 |
| Semi-detached | 20.74 | 300.73 | 248.78 | 82.43 | 631.94 |
| Detached | 24.14 | 350.03 | 302.54 | 97.89 | 750.46 |

| | Hours | Hours £ | Mat'ls £ | O & P £ | Total £ |
|---|---|---|---|---|---|
| **CAST IRON RAINWATER GOODS** | | | | | |
| **One-storey gable end** | | | | | |
| Terraced | 11.80 | 171.10 | 375.88 | 82.05 | 629.03 |
| Semi-detached | 14.10 | 204.45 | 391.14 | 89.34 | 684.93 |
| Detached | 15.70 | 227.65 | 415.72 | 96.51 | 739.88 |
| **Two-storey gable end** | | | | | |
| Terraced | 14.60 | 211.70 | 498.94 | 106.60 | 817.24 |
| Semi-detached | 16.90 | 245.05 | 514.20 | 113.89 | 873.14 |
| Detached | 18.50 | 268.25 | 538.78 | 121.05 | 928.08 |
| **Three-storey gable end** | | | | | |
| Terraced | 15.00 | 217.50 | 646.58 | 129.61 | 993.69 |
| Semi-detached | 19.30 | 279.85 | 661.84 | 141.25 | 1,082.94 |
| Detached | 21.90 | 317.55 | 686.42 | 150.60 | 1,154.57 |
| **One-storey hipped end** | | | | | |
| Terraced | 11.80 | 171.10 | 375.88 | 82.05 | 629.03 |
| Semi-detached | 17.10 | 247.95 | 462.43 | 106.56 | 816.94 |
| Detached | 22.40 | 324.80 | 622.90 | 142.16 | 1,089.86 |
| **Two-storey hipped end** | | | | | |
| Terraced | 15.60 | 226.20 | 498.94 | 108.77 | 833.91 |
| Semi-detached | 20.90 | 303.05 | 622.45 | 138.83 | 1,064.33 |
| Detached | 25.80 | 374.10 | 770.54 | 171.70 | 1,316.34 |
| **Three-storey hipped end** | | | | | |
| Terraced | 19.00 | 275.50 | 646.58 | 138.31 | 1,060.39 |
| Semi-detached | 24.60 | 356.70 | 782.40 | 170.87 | 1,309.97 |
| Detached | 29.80 | 432.10 | 942.76 | 206.23 | 1,581.09 |

| | Hours | Hours £ | Mat'ls £ | O & P £ | Total £ |
|---|---|---|---|---|---|
| **ALUMINIUM RAINWATER GOODS** | | | | | |
| **One-storey gable end** | | | | | |
| Terraced | 10.80 | 156.60 | 366.10 | 78.41 | 601.11 |
| Semi-detached | 14.10 | 204.45 | 378.70 | 87.47 | 670.62 |
| Detached | 15.70 | 227.65 | 400.62 | 94.24 | 722.51 |
| **Two-storey gable end** | | | | | |
| Terraced | 13.60 | 197.20 | 479.34 | 101.48 | 778.02 |
| Semi-detached | 16.90 | 245.05 | 491.64 | 110.50 | 847.19 |
| Detached | 18.50 | 268.25 | 513.96 | 117.33 | 899.54 |
| **Three-storey gable end** | | | | | |
| Terraced | 18.00 | 261.00 | 561.62 | 123.39 | 946.01 |
| Semi-detached | 20.30 | 294.35 | 627.30 | 138.25 | 1,059.90 |
| Detached | 21.90 | 317.55 | 649.22 | 145.02 | 1,111.79 |
| Terraced | 11.80 | 171.10 | 365.10 | 80.43 | 616.63 |
| Semi-detached | 17.10 | 247.95 | 508.23 | 113.43 | 869.61 |
| Detached | 22.40 | 324.80 | 631.74 | 143.48 | 1,100.02 |
| **Two-storey hipped end** | | | | | |
| Terraced | 14.60 | 211.70 | 478.44 | 103.52 | 793.66 |
| Semi-detached | 20.90 | 303.05 | 621.57 | 138.69 | 1,063.31 |
| Detached | 25.80 | 374.10 | 769.66 | 171.56 | 1,315.32 |
| **Three-storey hipped end** | | | | | |
| Terraced | 19.00 | 275.50 | 635.78 | 136.69 | 1,047.97 |
| Semi-detached | 24.60 | 356.70 | 771.78 | 169.27 | 1,297.75 |
| Detached | 29.80 | 432.10 | 932.16 | 204.64 | 1,568.90 |

| | Qty | Unit | Hours | Hours £ | Mat'ls £ | O & P £ | Total £ |
|---|---|---|---|---|---|---|---|
| **BATHROOMS** | | | | | | | |
| **Bathroom BCD with 1 bath with shower handset (B), 1 WC (C) and 1 lavatory basin (D)** | | | | | | | |
| **Stripping out** | | | | | | | |
| Take out sanitary fittings from existing bathroom, including supply pipes, overflows and wastes, remove debris and make good to floors and walls to receive new fittings and pipework | | Item | 6.00 | 87.00 | 15.00 | 15.30 | 117.30 |
| **Pipework** | | | | | | | |
| 15mm diameter copper supply pipe | 11.30 | m | 2.26 | 32.77 | 19.32 | 7.81 | 59.90 |
| extra over for | | | | | | | |
| elbow | 9 | nr | 1.62 | 23.49 | 4.50 | 4.20 | 32.19 |
| tee | 3 | nr | 0.66 | 9.57 | 2.10 | 1.75 | 13.42 |
| tap connector | 4 | nr | 0.36 | 5.22 | 7.21 | 1.86 | 14.29 |
| 19mm diameter MPVC-U overflow pipe | 0.50 | m | 0.10 | 1.45 | 0.86 | 0.35 | 2.66 |
| extra over for | | | | | | | |
| elbow | 1 | nr | 0.18 | 2.61 | 1.26 | 0.58 | 4.45 |
| straight connector | 1 | nr | 0.20 | 2.90 | 1.87 | 0.72 | 5.49 |
| 32mm diameter polypropylene waste pipe | 1 | m | 0.20 | 2.90 | 1.09 | 0.60 | 4.59 |
| **Carried forward** | | | 11.58 | 167.91 | 53.21 | 33.17 | 254.29 |

| | Qty | Unit | Hours | Hours £ | Mat'ls £ | O & P £ | Total £ |
|---|---|---|---|---|---|---|---|
| **Brought forward** | | | 11.58 | 167.91 | 53.21 | 33.17 | 254.29 |
| extra over for bend | 2 | nr | 0.36 | 5.22 | 1.84 | 1.06 | 8.12 |
| **Traps** | | | | | | | |
| 32mm diameter polypropylene P trap | 2 | nr | 0.60 | 8.70 | 8.40 | 2.57 | 19.67 |
| **Sanitary fittings** | | | | | | | |
| Acrylic reinforced bath 1700mm long complete with chromium-plated grip handles, 40mm waste fitting, overflow, chain, plug bath panels, 20mm chromium-plated taps and shower handset | 1 | nr | 2.75 | 39.88 | 251.21 | 43.66 | 334.75 |
| Vitreous china-low level WC suite comprising pan, plastic seat and cover, 9 litre cistern and brackets and plastic connecting pipe | 1 | nr | 2.45 | 35.53 | 204.56 | 36.01 | 276.10 |
| Vitreous china wash basin size 560 × 430mm, complete with 32mm waste fitting, overflow, chain, stay and plug, pair 13mm chromium-plated easy clean pillar taps, cast iron cantilever brackets and pedestal | 1 | nr | 2.20 | 31.90 | 131.51 | 24.51 | 187.92 |
| **Total** | | | 19.94 | 289.13 | 650.73 | 140.98 | 1,080.84 |

| | Qty | Unit | Hours | Hours £ | Mat'ls £ | O & P £ | Total £ |
|---|---|---|---|---|---|---|---|
| **Bathroom BCDF with 1 bath with shower handset (B), 1 WC (C), 1 lavatory basin (D) and 1 bidet (F)** | | | | | | | |
| **Stripping out** | | | | | | | |
| Take out sanitary fittings from existing bathroom, including supply pipes, overflows and wastes, remove debris and make good to floors and walls to receive new fittings and pipework | | Item | 6.00 | 87.00 | 15.00 | 15.30 | 117.30 |
| **Pipework** | | | | | | | |
| 15mm diameter copper supply pipe | 14.10 | m | 2.82 | 40.89 | 24.11 | 9.75 | 74.75 |
| extra over for | | | | | | | |
| elbow | 12 | nr | 2.16 | 31.32 | 6.00 | 5.60 | 42.92 |
| tee | 3 | nr | 0.66 | 9.57 | 2.10 | 1.75 | 13.42 |
| tap connector | 3 | nr | 0.54 | 7.83 | 10.83 | 2.80 | 21.46 |
| 19mm diameter MPVC-U overflow pipe | 1 | m | 0.20 | 2.90 | 1.72 | 0.69 | 5.31 |
| extra over for | | | | | | | |
| elbow | 2 | nr | 0.36 | 5.22 | 2.52 | 1.16 | 8.90 |
| straight connector | 2 | nr | 0.40 | 5.80 | 3.74 | 1.43 | 10.97 |
| 32mm diameter polypropylene waste pipe | 1 | m | 0.20 | 2.90 | 1.09 | 0.60 | 4.59 |
| extra over for | | | | | | | |
| bend | 2 | nr | 0.36 | 5.22 | 1.84 | 1.06 | 8.12 |
| **Carried forward** | | | 13.70 | 198.65 | 68.95 | 40.14 | 307.74 |

| | Qty | Unit | Hours | Hours £ | Mat'ls £ | O & P £ | Total £ |
|---|---|---|---|---|---|---|---|
| **Brought forward** | | | 13.70 | 198.65 | 68.95 | 40.14 | 307.74 |
| **Traps** | | | | | | | |
| 32mm diameter polypropylene P trap | 2 | nr | 0.60 | 8.70 | 8.40 | 2.57 | 19.67 |
| **Sanitary fittings** | | | | | | | |
| Acrylic reinforced bath 1700mm long complete with chromium-plated grip handles, 40mm waste fitting, overflow, chain, plug bath panels, 20mm chromium-plated taps and shower handset | 1 | nr | 2.75 | 39.88 | 251.21 | 43.66 | 334.75 |
| Vitreous china-low level WC suite comprising pan, plastic seat and cover, 9 litre cistern and brackets and plastic connecting pipe | 1 | nr | 2.45 | 35.53 | 204.56 | 36.01 | 276.10 |
| Vitreous china wash basin size 560 × 430mm, complete with 32mm waste fitting, overflow, chain, stay and plug, pair 13mm chromium-plated easy clean pillar taps, cast iron cantilever brackets and pedestal | 1 | nr | 2.20 | 31.90 | 131.51 | 24.51 | 187.92 |
| Bidet, free-standing with plain rim and chromium-plated monobloc spray and waste fittings complete | 1 | nr | 2.50 | 36.25 | 258.79 | 44.26 | 339.30 |
| **Total** | | | 24.20 | 350.90 | 923.42 | 191.15 | 1,465.47 |

| | Qty | Unit | Hours | Hours £ | Mat'ls £ | O & P £ | Total £ |
|---|---|---|---|---|---|---|---|
| **Bathroom BDF with 1 bath with shower handset (B), 1 lavatory basin (D) and 1 bidet (F)** | | | | | | | |
| **Stripping out** | | | | | | | |
| Take out sanitary fittings from existing bathroom, including supply pipes, overflows and wastes, remove debris and make good to floors and walls to receive new fittings and pipework | | Item | 6.00 | 87.00 | 15.00 | 15.30 | 117.30 |
| **Pipework** | | | | | | | |
| 15mm diameter copper supply pipe | 13 | m | 2.60 | 37.70 | 22.23 | 8.99 | 68.92 |
| extra over for | | | | | | | |
| elbow | 10 | nr | 1.80 | 26.10 | 5.00 | 4.67 | 35.77 |
| tee | 4 | nr | 0.88 | 12.76 | 2.80 | 2.33 | 17.89 |
| tap connector | 6 | nr | 1.08 | 15.66 | 19.86 | 5.33 | 40.85 |
| 19mm diameter MPVC-U overflow pipe | 1 | m | 0.20 | 2.90 | 1.72 | 0.69 | 5.31 |
| extra over for | | | | | | | |
| elbow | 2 | nr | 0.36 | 5.22 | 2.52 | 1.16 | 8.90 |
| straight connector | 2 | nr | 0.40 | 5.80 | 3.74 | 1.43 | 10.97 |
| 32mm diameter polypropylene waste pipe | 1 | m | 0.20 | 2.90 | 1.09 | 0.60 | 4.59 |
| extra over for | | | | | | | |
| bend | 2 | nr | 0.36 | 5.22 | 1.84 | 1.06 | 8.12 |
| **Carried forward** | | | 13.88 | 201.26 | 75.80 | 41.56 | 318.62 |

| | Qty | Unit | Hours | Hours £ | Mat'ls £ | O & P £ | Total £ |
|---|---|---|---|---|---|---|---|
| **Brought forward** | | | 13.88 | 201.26 | 75.80 | 41.56 | 318.62 |
| **Traps** | | | | | | | |
| 32mm diameter polypropylene P trap | 2 | nr | 0.60 | 8.70 | 8.20 | 2.54 | 19.44 |
| **Sanitary fittings** | | | | | | | |
| Acrylic reinforced bath 1700mm long complete with chromium-plated grip handles, 40mm waste fitting, overflow, chain, plug bath panels, 20mm chromium-plated taps and shower handset | 1 | nr | 2.75 | 39.88 | 251.21 | 43.66 | 334.75 |
| Vitreous china wash basin size 560 × 430mm, complete with 32mm waste fitting, overflow, chain, stay and plug, pair 13mm chromium-plated easy clean pillar taps, cast iron cantilever brackets and pedestal | 1 | nr | 2.20 | 31.90 | 131.50 | 24.51 | 187.91 |
| Bidet, free-standing with plain rim and chromium-plated monobloc spray and waste fittings complete | 1 | nr | 2.50 | 36.25 | 258.79 | 44.26 | 339.30 |
| **Total** | | | 21.93 | 317.99 | 725.50 | 156.52 | 1,200.01 |

| | Qty | Unit | Hours | Hours £ | Mat'ls £ | O & P £ | Total £ |
|---|---|---|---|---|---|---|---|
| **Bathroom BD with 1 bath with shower handset (B) and 1 lavatory basin (D)** | | | | | | | |
| **Stripping out** | | | | | | | |
| Take out sanitary fittings from existing bathroom, including supply pipes, overflows and wastes, remove debris and make good to floors and walls to receive new fittings and pipework | | Item | 6.00 | 87.00 | 15.00 | 15.30 | 117.30 |
| **Pipework** | | | | | | | |
| 15mm diameter copper supply pipe | 13.40 | m | 2.68 | 38.86 | 22.91 | 9.27 | 71.04 |
| extra over for | | | | | | | |
| elbow | 10 | nr | 1.80 | 26.10 | 5.00 | 4.67 | 35.77 |
| tee | 2 | nr | 0.44 | 6.38 | 1.40 | 1.17 | 8.95 |
| tap connector | 4 | nr | 0.72 | 10.44 | 14.44 | 3.73 | 28.61 |
| 19mm diameter MPVC-U overflow pipe | 0.50 | m | 0.10 | 1.45 | 0.86 | 0.35 | 2.66 |
| extra over for | | | | | | | |
| elbow | 2 | nr | 0.18 | 2.61 | 1.26 | 0.58 | 4.45 |
| straight connector | 2 | nr | 0.20 | 2.90 | 1.87 | 0.72 | 5.49 |
| 32mm diameter polypropylene waste pipe | 1 | m | 0.20 | 2.90 | 1.09 | 0.60 | 4.59 |
| extra over for | | | | | | | |
| bend | 2 | nr | 0.36 | 5.22 | 1.84 | 1.06 | 8.12 |
| **Carried forward** | | | 12.68 | 183.86 | 65.67 | 37.43 | 286.96 |

| | Qty | Unit | Hours | Hours £ | Mat'ls £ | O & P £ | Total £ |
|---|---|---|---|---|---|---|---|
| **Brought forward** | | | 12.68 | 183.86 | 65.67 | 37.43 | 286.96 |
| **Traps** | | | | | | | |
| 32mm diameter polypropylene P trap | 2 | nr | 0.60 | 8.70 | 8.20 | 2.54 | 19.44 |
| **Sanitary fittings** | | | | | | | |
| Acrylic reinforced bath 1700mm long complete with chromium-plated grip handles, 40mm waste fitting, overflow, chain, plug bath panels, 20mm chromium-plated taps and shower handset | 1 | nr | 2.75 | 39.88 | 251.21 | 43.66 | 334.75 |
| Vitreous china wash basin size 560 × 430mm, complete with 32mm waste fitting, overflow, chain, stay and plug, pair 13mm chromium-plated easy clean pillar taps, cast iron cantilever brackets and pedestal | 1 | nr | 2.20 | 31.90 | 131.51 | 24.51 | 187.92 |
| **Total** | | | 18.23 | 264.34 | 456.59 | 108.14 | 829.06 |

| | Qty | Unit | Hours | Hours £ | Mat'ls £ | O & P £ | Total £ |
|---|---|---|---|---|---|---|---|
| **Bathroom BCDE with 1 bath with shower handset (B), 1 WC (C), 1 lavatory basin (D) and 1 shower cubicle (E)** | | | | | | | |
| **Stripping out** | | | | | | | |
| Take out sanitary fittings from existing bathroom, including supply pipes, overflows and wastes, remove debris and make good to floors and walls to receive new fittings and pipework | | Item | 6.00 | 87.00 | 15.00 | 15.30 | 117.30 |
| **Pipework** | | | | | | | |
| 15mm diameter copper supply pipe | 16.30 | m | 3.26 | 47.27 | 27.87 | 11.27 | 86.41 |
| extra over for | | | | | | | |
| elbow | 11 | nr | 1.98 | 28.71 | 5.50 | 5.13 | 39.34 |
| tee | 5 | nr | 1.10 | 15.95 | 3.50 | 2.92 | 22.37 |
| tap connector | 4 | nr | 0.72 | 10.44 | 14.44 | 3.73 | 28.61 |
| 19mm diameter MPVC-U overflow pipe | 0.50 | m | 0.50 | 7.25 | 0.86 | 1.22 | 9.33 |
| extra over for | | | | | | | |
| elbow | 1 | nr | 0.18 | 2.61 | 1.26 | 0.58 | 4.45 |
| straight connector | 1 | nr | 0.20 | 2.90 | 1.87 | 0.72 | 5.49 |
| 32mm diameter polypropylene waste pipe | 1.50 | m | 0.30 | 4.35 | 1.63 | 0.90 | 6.88 |
| extra over for | | | | | | | |
| bend | 2 | nr | 0.36 | 5.22 | 1.84 | 1.06 | 8.12 |
| **Carried forward** | | | 14.60 | 211.70 | 73.77 | 42.82 | 328.29 |

| | Qty | Unit | Hours | Hours £ | Mat'ls £ | O & P £ | Total £ |
|---|---|---|---|---|---|---|---|
| **Brought forward** | | | 14.60 | 211.70 | 73.77 | 42.82 | 328.29 |
| **Traps** | | | | | | | |
| 32mm diameter polypropylene P trap | 3 | nr | 0.90 | 13.05 | 12.60 | 3.85 | 29.50 |
| **Sanitary fittings** | | | | | | | |
| Acrylic reinforced bath 1700mm long complete with chromium-plated grip handles, 40mm waste fitting, overflow, chain, plug bath panels, 20mm chromium-plated taps and shower handset | 1 | nr | 2.75 | 39.88 | 251.21 | 43.66 | 334.75 |
| Vitreous china-low level WC suite comprising pan, plastic seat and cover, 9 litre cistern and brackets and plastic connecting pipe | 1 | nr | 2.45 | 35.53 | 204.56 | 36.01 | 276.10 |
| Vitreous china wash basin size 560 × 430mm, complete with 32mm waste fitting, overflow, chain, stay and plug, pair 13mm chromium-plated easy clean pillar taps, cast iron cantilever brackets and pedestal | 1 | nr | 2.20 | 31.90 | 131.51 | 24.51 | 187.92 |
| **Carried forward** | | | 22.90 | 332.05 | 673.65 | 150.85 | 1,156.55 |

| | Qty | Unit | Hours | Hours £ | Mat'ls £ | O & P £ | Total £ |
|---|---|---|---|---|---|---|---|
| **Brought forward** | | | 22.90 | 332.05 | 673.65 | 150.85 | 1,156.55 |
| Shower cubicle size 788 × 842 × 2115mm high, anodised aluminium frame and safety glass, complete with 8.15kW instant electric shower with adjustable spray handset, slide bar and hose, soap dish and chromium-plated waste fittings | 1 | nr | 3.80 | 55.10 | 689.48 | 111.69 | 856.27 |
| **Total** | | | 26.70 | 387.15 | 1,363.13 | 262.54 | 2,012.82 |

| | Qty | Unit | Hours | Hours £ | Mat'ls £ | O & P £ | Total £ |
|---|---|---|---|---|---|---|---|
| **Bathroom BCDEF with 1 bath with shower handset (B), 1 WC (C), 1 lavatory basin (D), 1 shower cubicle (E) and 1 bidet (F)** | | | | | | | |
| **Stripping out** | | | | | | | |
| Take out sanitary fittings from existing bathroom, including supply pipes, overflows and wastes, remove debris and make good to floors and walls to receive new fittings and pipework | | Item | 6.00 | 87.00 | 15.00 | 15.30 | 117.30 |
| **Pipework** | | | | | | | |
| 15mm diameter copper supply pipe | 19.90 | m | 3.98 | 57.71 | 34.02 | 13.76 | 105.49 |
| **Carried forward** | | | 9.98 | 144.71 | 49.02 | 29.06 | 222.79 |

| | Qty | Unit | Hours | Hours £ | Mat'ls £ | O & P £ | Total £ |
|---|---|---|---|---|---|---|---|
| **Brought forward** | | | 9.98 | 144.71 | 49.02 | 29.06 | 222.79 |
| extra over for | | | | | | | |
| elbow | 14 | nr | 2.52 | 36.54 | 7.00 | 6.53 | 50.07 |
| tee | 7 | nr | 1.54 | 22.33 | 4.90 | 4.08 | 31.31 |
| tap connector | 6 | nr | 0.72 | 10.44 | 21.66 | 4.82 | 36.92 |
| 19mm diameter MPVC-U overflow pipe | 1 | m | 0.20 | 2.90 | 1.72 | 0.69 | 5.31 |
| extra over for | | | | | | | |
| elbow | 2 | nr | 0.36 | 5.22 | 2.52 | 1.16 | 8.90 |
| straight connector | 2 | nr | 0.40 | 5.80 | 3.74 | 1.43 | 10.97 |
| 32mm diameter polypropylene waste pipe | 1.50 | m | 0.30 | 4.35 | 1.63 | 0.90 | 6.88 |
| extra over for | | | | | | | |
| bend | 2 | nr | 0.36 | 5.22 | 1.84 | 1.06 | 8.12 |
| **Traps** | | | | | | | |
| 32mm diameter polypropylene P trap | 3 | nr | 0.90 | 13.05 | 12.60 | 3.85 | 29.50 |
| **Sanitary fittings** | | | | | | | |
| Acrylic reinforced bath 1700mm long complete with chromium-plated grip handles, 40mm waste fitting, overflow, chain, plug bath panels, 20mm chromium-plated taps and shower handset | 1 | nr | 2.75 | 39.88 | 251.21 | 43.66 | 334.75 |
| **Carried forward** | | | 20.03 | 290.44 | 357.84 | 97.24 | 745.52 |

| | Qty | Unit | Hours | Hours £ | Mat'ls £ | O & P £ | Total £ |
|---|---|---|---|---|---|---|---|
| **Brought forward** | | | 20.03 | 290.44 | 357.84 | 97.24 | 745.52 |
| Vitreous china-low level WC suite comprising pan, plastic seat and cover, 9 litre cistern and brackets and plastic connecting pipe | 1 | nr | 2.45 | 35.53 | 204.56 | 36.01 | 276.10 |
| Vitreous china wash basin size 560 × 430mm, complete with 32mm waste fitting, overflow, chain, stay and plug, pair 13mm chromium-plated easy clean pillar taps, cast iron cantilever brackets and pedestal | 1 | nr | 2.20 | 31.90 | 131.51 | 24.51 | 187.92 |
| Shower cubicle size 788 × 842 × 2115mm high, anodised aluminium frame and safety glass, complete with 8.15kW instant electric shower with adjustable spray handset, slide bar and hose, soap dish and chromium-plated waste fittings | 1 | nr | 3.80 | 55.10 | 689.48 | 111.69 | 856.27 |
| Bidet, free-standing with plain rim and chromium-plated monobloc spray and waste fittings complete | 1 | nr | 2.50 | 36.25 | 258.79 | 44.26 | 339.30 |
| **Total** | | | 30.98 | 449.22 | 1,642.18 | 313.71 | 2,405.10 |

| | Qty | Unit | Hours | Hours £ | Mat'ls £ | O & P £ | Total £ |
|---|---|---|---|---|---|---|---|
| **Bathroom BDEF with 1 bath with shower handset (B), 1 lavatory basin (D), 1 shower cubicle (E) and 1 bidet (F)** | | | | | | | |
| **Stripping out** | | | | | | | |
| Take out sanitary fittings from existing bathroom, including supply pipes, overflows and wastes, remove debris and make good to floors and walls to receive new fittings and pipework | | Item | 6.00 | 87.00 | 15.00 | 15.30 | 117.30 |
| **Pipework** | | | | | | | |
| 15mm diameter copper supply pipe | 15.80 | m | 3.16 | 45.82 | 27.02 | 10.93 | 83.77 |
| extra over for | | | | | | | |
| elbow | 12 | nr | 2.16 | 31.32 | 6.00 | 5.60 | 42.92 |
| tee | 6 | nr | 1.32 | 19.14 | 4.20 | 3.50 | 26.84 |
| tap connector | 6 | nr | 1.08 | 15.66 | 21.66 | 5.60 | 42.92 |
| 19mm diameter MPVC-U overflow pipe | 1 | m | 0.20 | 2.90 | 1.72 | 0.69 | 5.31 |
| extra over for | | | | | | | |
| elbow | 2 | nr | 0.36 | 5.22 | 2.52 | 1.16 | 8.90 |
| straight connector | 2 | nr | 0.40 | 5.80 | 3.74 | 1.43 | 10.97 |
| 32mm diameter polypropylene waste pipe | 1.50 | m | 0.30 | 4.35 | 1.63 | 0.90 | 6.88 |
| extra over for | | | | | | | |
| bend | 2 | nr | 0.36 | 5.22 | 1.84 | 1.06 | 8.12 |
| **Carried forward** | | | 15.34 | 222.43 | 85.33 | 46.16 | 353.92 |

| | Qty | Unit | Hours | Hours £ | Mat'ls £ | O & P £ | Total £ |
|---|---|---|---|---|---|---|---|
| **Brought forward** | | | 15.34 | 222.43 | 85.33 | 46.16 | 353.92 |
| **Traps** | | | | | | | |
| 32mm diameter polypropylene P trap | 3 | nr | 0.90 | 13.05 | 12.60 | 3.85 | 29.50 |
| **Sanitary fittings** | | | | | | | |
| Acrylic reinforced bath 1700mm long complete with chromium-plated grip handles, 40mm waste fitting, overflow, chain, plug bath panels, 20mm chromium-plated taps and shower handset | 1 | nr | 2.75 | 39.88 | 251.21 | 43.66 | 334.75 |
| Vitreous china wash basin size 560 × 430mm, complete with 32mm waste fitting, overflow, chain, stay and plug, pair 13mm chromium-plated easy clean pillar taps, cast iron cantilever brackets and pedestal | 1 | nr | 2.20 | 31.90 | 131.51 | 24.51 | 187.92 |
| Shower cubicle size 788 × 842 × 2115mm high, anodised aluminium frame and safety glass, complete with 8.15kW instant electric shower with adjustable spray handset, slide bar and hose, soap dish and chromium-plated waste fittings | 1 | nr | 3.80 | 55.10 | 689.48 | 111.69 | 856.27 |
| **Carried forward** | | | 24.99 | 362.36 | 1,170.13 | 229.87 | 1,762.35 |

| | Qty | Unit | Hours | Hours £ | Mat'ls £ | O & P £ | Total £ |
|---|---|---|---|---|---|---|---|
| **Brought forward** | | | 24.99 | 362.36 | 1,170.13 | 229.87 | 1,762.35 |
| Bidet, free-standing with plain rim and chromium-plated monobloc spray and waste fittings complete | 1 | nr | 2.50 | 36.25 | 258.79 | 44.26 | 339.30 |
| **Total** | | | 27.49 | 398.61 | 1,428.92 | 274.13 | 2,101.65 |

| | Qty | Unit | Hours | Hours £ | Mat'ls £ | O & P £ | Total £ |
|---|---|---|---|---|---|---|---|
| **Bathroom BDE with 1 bath with shower handset (B), 1 lavatory basin (D) and 1 shower cubicle (E)** | | | | | | | |
| **Stripping out** | | | | | | | |
| Take out sanitary fittings from existing bathroom, including supply pipes, overflows and wastes, remove debris and make good to floors and walls to receive new fittings and pipework | | Item | 6.00 | 87.00 | 15.00 | 15.30 | 117.30 |
| **Pipework** | | | | | | | |
| 15mm diameter copper supply pipe | 15.80 | m | 3.16 | 45.82 | 27.02 | 10.93 | 83.77 |
| extra over for | | | | | | | |
| elbow | 10 | nr | 1.80 | 26.10 | 5.00 | 4.67 | 35.77 |
| tee | 4 | nr | 0.88 | 12.76 | 2.80 | 2.33 | 17.89 |
| tap connector | 6 | nr | 1.08 | 15.66 | 21.66 | 5.60 | 42.92 |
| **Carried forward** | | | 12.92 | 187.34 | 71.48 | 38.82 | 297.64 |

| | Qty | Unit | Hours | Hours £ | Mat'ls £ | O & P £ | Total £ |
|---|---|---|---|---|---|---|---|
| **Brought forward** | | | 12.92 | 187.34 | 71.48 | 38.82 | 297.64 |
| 19mm diameter MPVC-U overflow pipe | 0.50 | m | 0.10 | 1.45 | 0.86 | 0.35 | 2.66 |
| extra over for | | | | | | | |
| elbow | 1 | nr | 0.18 | 2.61 | 0.26 | 0.43 | 3.30 |
| straight connector | 1 | nr | 0.20 | 2.90 | 1.87 | 0.72 | 5.49 |
| 32mm diameter polypropylene waste pipe | 1.00 | m | 0.20 | 2.90 | 1.09 | 0.60 | 4.59 |
| extra over for | | | | | | | |
| bend | 2 | nr | 0.36 | 5.22 | 1.84 | 1.06 | 8.12 |
| **Traps** | | | | | | | |
| 32mm diameter polypropylene P trap | 3 | nr | 0.90 | 13.05 | 12.60 | 3.85 | 29.50 |
| **Sanitary fittings** | | | | | | | |
| Acrylic reinforced bath 1700mm long complete with chromium-plated grip handles, 40mm waste fitting, overflow, chain, plug bath panels, 20mm chromium-plated taps and shower handset | 1 | nr | 2.75 | 39.88 | 251.21 | 43.66 | 334.75 |
| Vitreous china wash basin size 560 × 430mm, complete with 32mm waste fitting, overflow, chain, stay and plug, pair 13mm chromium-plated easy clean pillar taps, cast iron cantilever brackets and pedestal | 1 | nr | 2.20 | 31.90 | 131.51 | 24.51 | 187.92 |
| **Carried forward** | | | 19.81 | 287.25 | 472.72 | 113.99 | 873.96 |

| | Qty | Unit | Hours | Hours £ | Mat'ls £ | O & P £ | Total £ |
|---|---|---|---|---|---|---|---|
| **Brought forward** | | | 19.81 | 287.25 | 472.72 | 113.99 | 873.96 |
| Shower cubicle size 788 × 842 × 2115mm high, anodised aluminium frame and safety glass, complete with 8.15kW instant electric shower with adjustable spray handset, slide bar and hose, soap dish and chromium-plated waste fittings | 1 | nr | 3.80 | 55.10 | 689.48 | 111.69 | 856.27 |
| **Total** | | | 23.61 | 342.35 | 1,162.20 | 225.68 | 1,730.23 |

| | Qty | Unit | Hours | Hours £ | Mat'ls £ | O & P £ | Total £ |
|---|---|---|---|---|---|---|---|
| **Bathroom ACD with 1 bath (A), 1 WC (C) and 1 lavatory basin (D)** | | | | | | | |
| **Stripping out** | | | | | | | |
| Take out sanitary fittings from existing bathroom, including supply pipes, overflows and wastes, remove debris and make good to floors and walls to receive new fittings and pipework | | Item | 6.00 | 87.00 | 15.00 | 15.30 | 117.30 |
| **Pipework** | | | | | | | |
| 15mm diameter copper supply pipe | 11.30 | m | 2.26 | 32.77 | 19.32 | 7.81 | 59.90 |
| **Carried forward** | | | 8.26 | 119.77 | 34.32 | 23.11 | 177.20 |

| | Qty | Unit | Hours | Hours £ | Mat'ls £ | O & P £ | Total £ |
|---|---|---|---|---|---|---|---|
| **Brought forward** | | | 8.26 | 119.77 | 34.32 | 23.11 | 177.20 |
| extra over for | | | | | | | |
| elbow | 9 | nr | 1.80 | 26.10 | 5.00 | 4.67 | 35.77 |
| tee | 3 | nr | 0.88 | 12.76 | 2.80 | 2.33 | 17.89 |
| tap connector | 4 | nr | 0.72 | 10.44 | 14.44 | 3.73 | 28.61 |
| 19mm diameter MPVC-U overflow pipe | 0.50 | m | 0.10 | 1.45 | 0.86 | 0.35 | 2.66 |
| extra over for | | | | | | | |
| elbow | 1 | nr | 0.18 | 2.61 | 1.26 | 0.58 | 4.45 |
| straight connector | 1 | nr | 0.20 | 2.90 | 1.87 | 0.72 | 5.49 |
| 32mm diameter polypropylene waste pipe | 1 | m | 0.20 | 2.90 | 1.09 | 0.60 | 4.59 |
| extra over for | | | | | | | |
| bend | 1 | nr | 0.36 | 5.22 | 1.84 | 1.06 | 8.12 |
| **Traps** | | | | | | | |
| 32mm diameter polypropylene P trap | 1 | nr | 0.60 | 8.70 | 8.40 | 2.57 | 19.67 |
| **Sanitary fittings** | | | | | | | |
| Acrylic reinforced bath 1700mm long complete with chromium-plated grip handles, 40mm waste fitting, overflow, chain, plug bath panels, 20mm chromium-plated taps | 1 | nr | 2.75 | 39.88 | 171.24 | 31.67 | 242.78 |
| **Carried forward** | | | 16.05 | 232.73 | 243.12 | 71.37 | 547.22 |

| | Qty | Unit | Hours | Hours £ | Mat'ls £ | O & P £ | Total £ |
|---|---|---|---|---|---|---|---|
| **Brought forward** | | | 16.05 | 232.73 | 243.12 | 71.37 | 547.22 |
| Vitreous china-low level WC suite comprising pan, plastic seat and cover, 9 litre cistern and brackets and plastic connecting pipe | 1 | nr | 2.45 | 35.53 | 204.56 | 36.01 | 276.10 |
| Vitreous china wash basin size 560 × 430mm, complete with 32mm waste fitting, overflow, chain, stay and plug, pair 13mm chromium-plated easy clean pillar taps, cast iron cantilever brackets and pedestal | 1 | nr | 2.20 | 31.90 | 131.51 | 24.51 | 187.92 |
| **Total** | | | 20.70 | 300.16 | 579.19 | 131.89 | 1,011.24 |

| | Qty | Unit | Hours | Hours £ | Mat'ls £ | O & P £ | Total £ |
|---|---|---|---|---|---|---|---|
| **Bathroom ACDF with 1 bath (A), 1 WC (C), 1 lavatory basin (D) and 1 bidet (F)** | | | | | | | |
| **Stripping out** | | | | | | | |
| Take out sanitary fittings from existing bathroom, including supply pipes, overflows and wastes, remove debris and make good to floors and walls to receive new fittings and pipework | | Item | 6.00 | 87.00 | 15.00 | 15.30 | 117.30 |
| **Carried forward** | | | 6.00 | 87.00 | 15.00 | 15.30 | 117.30 |

| | Qty | Unit | Hours | Hours £ | Mat'ls £ | O & P £ | Total £ |
|---|---|---|---|---|---|---|---|
| **Brought forward** | | | 6.00 | 87.00 | 15.00 | 15.30 | 117.30 |
| **Pipework** | | | | | | | |
| 15mm diameter copper supply pipe | 14.10 | m | 2.82 | 40.89 | 21.15 | 9.31 | 71.35 |
| extra over for | | | | | | | |
| elbow | 12 | nr | 2.16 | 31.32 | 6.00 | 5.60 | 42.92 |
| tee | 4 | nr | 0.66 | 9.57 | 2.10 | 1.75 | 13.42 |
| tap connector | 3 | nr | 0.54 | 7.83 | 10.83 | 2.80 | 21.46 |
| 19mm diameter MPVC-U overflow pipe | 1 | m | 0.20 | 2.90 | 1.72 | 0.69 | 5.31 |
| extra over for | | | | | | | |
| elbow | 2 | nr | 0.36 | 5.22 | 2.52 | 1.16 | 8.90 |
| straight connector | 2 | nr | 0.40 | 5.80 | 3.74 | 1.43 | 10.97 |
| 32mm diameter polypropylene waste pipe | 1 | m | 0.20 | 2.90 | 1.09 | 0.60 | 4.59 |
| extra over for | | | | | | | |
| bend | 2 | nr | 0.36 | 5.22 | 1.84 | 1.06 | 8.12 |
| **Traps** | | | | | | | |
| 32mm diameter polypropylene P trap | 2 | nr | 0.60 | 8.70 | 8.40 | 2.57 | 19.67 |
| **Sanitary fittings** | | | | | | | |
| Acrylic reinforced bath 1700mm long complete with chromium-plated grip handles, 40mm waste fitting, overflow, chain, plug bath panels, 20mm chromium-plated taps | 1 | nr | 2.75 | 39.88 | 251.21 | 43.66 | 334.75 |
| **Carried forward** | | | 17.05 | 247.23 | 325.60 | 85.92 | 658.75 |

| | Qty | Unit | Hours | Hours £ | Mat'ls £ | O & P £ | Total £ |
|---|---|---|---|---|---|---|---|
| **Brought forward** | | | 17.05 | 247.23 | 325.60 | 85.92 | 658.75 |
| Vitreous china-low level WC suite comprising pan, plastic seat and cover, 9 litre cistern and brackets and plastic connecting pipe | 1 | nr | 2.45 | 35.53 | 204.56 | 36.01 | 276.10 |
| Vitreous china wash basin size 560 × 430mm, complete with 32mm waste fitting, overflow, chain, stay and plug, pair 13mm chromium-plated easy clean pillar taps, cast iron cantilever brackets and pedestal | 1 | nr | 2.20 | 31.90 | 131.51 | 24.51 | 187.92 |
| Bidet, free-standing with plain rim and chromium-plated monobloc spray and waste fittings complete | 1 | nr | 2.50 | 36.25 | 258.79 | 44.26 | 339.30 |
| **Total** | | | 24.20 | 350.91 | 920.46 | 190.70 | 1,462.07 |

| | Qty | Unit | Hours | Hours £ | Mat'ls £ | O & P £ | Total £ |
|---|---|---|---|---|---|---|---|
| **Bathroom ADF with 1 bath (A), 1 lavatory basin (D) and 1 bidet (F)** | | | | | | | |
| **Stripping out** | | | | | | | |
| Take out sanitary fittings from existing bathroom, including supply pipes, overflows and wastes, remove debris and make good to floors and walls to receive new fittings and pipework | | Item | 6.00 | 87.00 | 15.00 | 15.30 | 117.30 |
| **Pipework** | | | | | | | |
| 15mm diameter copper supply pipe | 13 | m | 2.60 | 37.70 | 22.33 | 9.00 | 69.03 |
| extra over for | | | | | | | |
| elbow | 10 | nr | 1.80 | 26.10 | 5.00 | 4.67 | 35.77 |
| tee | 4 | nr | 0.88 | 12.76 | 2.80 | 2.33 | 17.89 |
| tap connector | 6 | nr | 1.08 | 15.66 | 19.86 | 5.33 | 40.85 |
| 19mm diameter MPVC-U overflow pipe | 1 | m | 0.25 | 3.63 | 2.38 | 0.90 | 6.91 |
| extra over for | | | | | | | |
| elbow | 2 | nr | 0.20 | 2.90 | 1.72 | 0.69 | 5.31 |
| straight connector | 2 | nr | 0.40 | 5.80 | 2.62 | 1.26 | 9.68 |
| 32mm diameter polypropylene waste pipe | 1 | m | 0.36 | 5.22 | 2.52 | 1.16 | 8.90 |
| extra over for | | | | | | | |
| bend | 2 | nr | 0.40 | 5.80 | 3.74 | 1.43 | 10.97 |
| **Carried forward** | | | 13.97 | 202.57 | 77.97 | 42.08 | 322.62 |

| | Qty | Unit | Hours | Hours £ | Mat'ls £ | O & P £ | Total £ |
|---|---|---|---|---|---|---|---|
| **Brought forward** | | | 13.97 | 202.57 | 77.97 | 42.08 | 322.62 |
| **Traps** | | | | | | | |
| 32mm diameter polypropylene P trap | 2 | nr | 0.60 | 8.70 | 8.20 | 2.54 | 19.44 |
| **Sanitary fittings** | | | | | | | |
| Acrylic reinforced bath 1700mm long complete with chromium-plated grip handles, 40mm waste fitting, overflow, chain, plug bath panels, 20mm chromium-plated taps | 1 | nr | 2.75 | 39.88 | 171.74 | 31.74 | 243.36 |
| Vitreous china wash basin size 560 × 430mm, complete with 32mm waste fitting, overflow, chain, stay and plug, pair 13mm chromium-plated easy clean pillar taps, cast iron cantilever brackets and pedestal | 1 | nr | 2.45 | 35.53 | 204.56 | 36.01 | 276.10 |
| Bidet, free-standing with plain rim and chromium-plated monobloc spray and waste fittings complete | 1 | nr | 2.20 | 31.90 | 131.51 | 24.51 | 187.92 |
| **Total** | | | 21.97 | 318.57 | 593.98 | 136.88 | 1,049.43 |

| | Qty | Unit | Hours | Hours £ | Mat'ls £ | O & P £ | Total £ |
|---|---|---|---|---|---|---|---|
| **Bathroom AD with 1 bath (A) and 1 lavatory basin (D)** | | | | | | | |
| **Stripping out** | | | | | | | |
| Take out sanitary fittings from existing bathroom, including supply pipes, overflows and wastes, remove debris and make good to floors and walls to receive new fittings and pipework | | Item | 6.00 | 87.00 | 15.00 | 15.30 | 117.30 |
| **Pipework** | | | | | | | |
| 15mm diameter copper supply pipe | 13.40 | m | 2.68 | 38.86 | 22.91 | 9.27 | 71.04 |
| extra over for | | | | | | | |
| elbow | 10 | nr | 1.80 | 26.10 | 5.00 | 4.67 | 35.77 |
| tee | 2 | nr | 0.44 | 6.38 | 1.40 | 1.17 | 8.95 |
| tap connector | 4 | nr | 0.72 | 10.44 | 14.44 | 3.73 | 28.61 |
| 19mm diameter MPVC-U overflow pipe | 0.50 | m | 0.10 | 1.45 | 0.86 | 0.35 | 2.66 |
| extra over for | | | | | | | |
| elbow | 2 | nr | 0.18 | 2.61 | 1.26 | 0.58 | 4.45 |
| straight connector | 2 | nr | 0.20 | 2.90 | 1.87 | 0.72 | 5.49 |
| 32mm diameter polypropylene waste pipe | 1 | m | 0.20 | 2.90 | 1.09 | 0.60 | 4.59 |
| extra over for | | | | | | | |
| bend | 2 | nr | 0.36 | 5.22 | 1.84 | 1.06 | 8.12 |
| **Carried forward** | | | 12.68 | 183.86 | 65.67 | 37.43 | 286.96 |

| | Qty | Unit | Hours | Hours £ | Mat'ls £ | O & P £ | Total £ |
|---|---|---|---|---|---|---|---|
| **Brought forward** | | | 12.68 | 183.86 | 65.67 | 37.43 | 286.96 |
| **Traps** | | | | | | | |
| 32mm diameter polypropylene P trap | 2 | nr | 0.60 | 8.70 | 8.20 | 2.54 | 19.44 |
| **Sanitary fittings** | | | | | | | |
| Acrylic reinforced bath 1700mm long complete with chromium-plated grip handles, 40mm waste fitting, overflow, chain, plug bath panels, 20mm chromium-plated taps | 1 | nr | 2.75 | 39.88 | 171.24 | 31.67 | 242.78 |
| Vitreous china wash basin size 560 × 430mm, complete with 32mm waste fitting, overflow, chain, stay and plug, pair 13mm chromium-plated easy clean pillar taps, cast iron cantilever brackets and pedestal | 1 | nr | 2.20 | 31.90 | 131.51 | 24.51 | 187.92 |
| **Total** | | | 18.23 | 264.34 | 376.62 | 96.14 | 737.10 |

| | Qty | Unit | Hours | Hours £ | Mat'ls £ | O & P £ | Total £ |
|---|---|---|---|---|---|---|---|
| **Bathroom ACDE with 1 bath (A), 1 WC (C), 1 lavatory basin (D) and 1 shower cubicle (E)** | | | | | | | |
| **Stripping out** | | | | | | | |
| Take out sanitary fittings from existing bathroom, including supply pipes, overflows and wastes, remove debris and make good to floors and walls to receive new fittings and pipework | | Item | 6.00 | 87.00 | 15.00 | 15.30 | 117.30 |
| **Pipework** | | | | | | | |
| 15mm diameter copper supply pipe | 16.30 | m | 3.26 | 47.27 | 27.87 | 11.27 | 86.41 |
| extra over for | | | | | | | |
| elbow | 11 | nr | 1.98 | 28.71 | 5.50 | 5.13 | 39.34 |
| tee | 5 | nr | 1.10 | 15.95 | 3.50 | 2.92 | 22.37 |
| tap connector | 4 | nr | 0.72 | 10.44 | 14.44 | 3.73 | 28.61 |
| 19mm diameter MPVC-U overflow pipe | 0.50 | m | 0.10 | 1.45 | 0.86 | 0.35 | 2.66 |
| extra over for | | | | | | | |
| elbow | 1 | nr | 0.18 | 2.61 | 1.26 | 0.58 | 4.45 |
| straight connector | 1 | nr | 0.20 | 2.90 | 1.87 | 0.72 | 5.49 |
| 32mm diameter polypropylene waste pipe | 1.50 | m | 0.30 | 4.35 | 1.63 | 0.90 | 6.88 |
| extra over for | | | | | | | |
| bend | 2 | nr | 0.36 | 5.22 | 1.84 | 1.06 | 8.12 |
| **Carried forward** | | | 14.20 | 200.68 | 71.93 | 40.89 | 313.50 |

| | Qty | Unit | Hours | Hours £ | Mat'ls £ | O & P £ | Total £ |
|---|---|---|---|---|---|---|---|
| **Brought forward** | | | 14.20 | 200.68 | 71.93 | 40.89 | 313.50 |
| **Traps** | | | | | | | |
| 32mm diameter polypropylene P trap | 3 | nr | 0.90 | 13.05 | 12.60 | 3.85 | 29.50 |
| **Sanitary fittings** | | | | | | | |
| Acrylic reinforced bath 1700mm long complete with chromium-plated grip handles, 40mm waste fitting, overflow, chain, plug bath panels, 20mm chromium-plated taps | 1 | nr | 2.75 | 39.88 | 171.24 | 31.67 | 242.78 |
| Vitreous china-low level WC suite comprising pan, plastic seat and cover, 9 litre cistern and brackets and plastic connecting pipe | 1 | nr | 2.45 | 35.53 | 204.56 | 36.01 | 276.10 |
| Vitreous china wash basin size 560 × 430mm, complete with 32mm waste fitting, overflow, chain, stay and plug, pair 13mm chromium-plated easy clean pillar taps, cast iron cantilever brackets and pedestal | 1 | nr | 2.20 | 31.90 | 131.51 | 24.51 | 187.92 |
| **Carried forward** | | | 22.50 | 321.03 | 591.84 | 136.93 | 1,049.80 |

| | Qty | Unit | Hours | Hours £ | Mat'ls £ | O & P £ | Total £ |
|---|---|---|---|---|---|---|---|
| **Brought forward** | | | 22.50 | 321.03 | 591.84 | 136.93 | 1,049.80 |
| Shower cubicle size 788 × 842 × 2115mm high, anodised aluminium frame and safety glass, complete with 8.15kW instant electric shower with adjustable spray handset, slide bar and hose, soap dish and chromium-plated waste fittings | 1 | nr | 3.80 | 55.10 | 689.48 | 111.69 | 856.27 |
| **Total** | | | 26.30 | 376.13 | 1,281.32 | 248.62 | 1,906.07 |

| | Qty | Unit | Hours | Hours £ | Mat'ls £ | O & P £ | Total £ |
|---|---|---|---|---|---|---|---|
| **Bathroom ACDEF with 1 bath (A), 1 WC (C), 1 lavatory basin (D), 1 shower cubicle (E) and 1 bidet (F)** | | | | | | | |
| **Stripping out** | | | | | | | |
| Take out sanitary fittings from existing bathroom, including supply pipes, overflows and wastes, remove debris and make good to floors and walls to receive new fittings and pipework | | Item | 6.00 | 87.00 | 15.00 | 15.30 | 117.30 |
| **Pipework** | | | | | | | |
| 15mm diameter copper supply pipe | 19.90 | m | 3.98 | 57.71 | 34.02 | 13.76 | 105.49 |
| **Carried forward** | | | 9.98 | 144.71 | 49.02 | 29.06 | 222.79 |

| | Qty | Unit | Hours | Hours £ | Mat'ls £ | O & P £ | Total £ |
|---|---|---|---|---|---|---|---|
| **Brought forward** | | | 9.98 | 144.71 | 49.02 | 29.06 | 222.79 |
| extra over for | | | | | | | |
| elbow | 14 | nr | 2.52 | 36.54 | 7.00 | 6.53 | 50.07 |
| tee | 5 | nr | 1.54 | 22.33 | 4.90 | 4.08 | 31.31 |
| tap connector | 4 | nr | 1.08 | 15.66 | 21.66 | 5.60 | 42.92 |
| 19mm diameter MPVC-U overflow pipe | 1 | m | 0.20 | 2.90 | 1.72 | 0.69 | 5.31 |
| extra over for | | | | | | | |
| elbow | 1 | nr | 0.36 | 5.22 | 2.52 | 1.16 | 8.90 |
| straight connector | 1 | nr | 0.40 | 5.80 | 3.74 | 1.43 | 10.97 |
| 32mm diameter polypropylene waste pipe | 1.50 | m | 0.30 | 4.35 | 1.63 | 0.90 | 6.88 |
| extra over for | | | | | | | |
| bend | 2 | nr | 0.36 | 5.22 | 1.84 | 1.06 | 8.12 |
| **Traps** | | | | | | | |
| 32mm diameter polypropylene P trap | 3 | nr | 0.90 | 13.05 | 12.60 | 3.85 | 29.50 |
| **Sanitary fittings** | | | | | | | |
| Acrylic reinforced bath 1700mm long complete with chromium-plated grip handles, 40mm waste fitting, overflow, chain, plug bath panels, 20mm chromium-plated taps | 1 | nr | 2.75 | 39.88 | 171.24 | 31.67 | 242.78 |
| **Carried forward** | | | 20.39 | 295.66 | 277.87 | 86.03 | 659.55 |

| | Qty | Unit | Hours | Hours £ | Mat'ls £ | O & P £ | Total £ |
|---|---|---|---|---|---|---|---|
| **Brought forward** | | | 20.39 | 295.66 | 277.87 | 86.03 | 659.55 |
| Vitreous china-low level WC suite comprising pan, plastic seat and cover, 9 litre cistern and brackets and plastic connecting pipe | 1 | nr | 2.45 | 35.53 | 204.56 | 36.01 | 276.10 |
| Vitreous china wash basin size 560 × 430mm, complete with 32mm waste fitting, overflow, chain, stay and plug, pair 13mm chromium-plated easy clean pillar taps, cast iron cantilever brackets and pedestal | 1 | nr | 2.20 | 31.90 | 131.51 | 24.51 | 187.92 |
| Shower cubicle size 788 × 842 × 2115mm high, anodised aluminium frame and safety glass, complete with 8.15kW instant electric shower with adjustable spray handset, slide bar and hose, soap dish and chromium-plated waste fittings | 1 | nr | 3.80 | 55.10 | 689.48 | 111.69 | 856.27 |
| Bidet, free-standing with plain rim and chromium-plated monobloc spray and waste fittings complete | 1 | nr | 2.50 | 36.25 | 258.79 | 44.26 | 339.30 |
| **Total** | | | 31.34 | 454.44 | 1,562.21 | 302.50 | 2,319.13 |

| | Qty | Unit | Hours | Hours £ | Mat'ls £ | O & P £ | Total £ |
|---|---|---|---|---|---|---|---|
| **Bathroom ADEF with 1 bath (A), 1 lavatory basin (D), 1 shower cubicle (E) and 1 bidet (F)** | | | | | | | |
| **Stripping out** | | | | | | | |
| Take out sanitary fittings from existing bathroom, including supply pipes, overflows and wastes, remove debris and make good to floors and walls to receive new fittings and pipework | | Item | 6.00 | 87.00 | 15.00 | 15.30 | 117.30 |
| **Pipework** | | | | | | | |
| 15mm diameter copper supply pipe | 15.80 | m | 3.16 | 45.82 | 25.28 | 10.67 | 81.77 |
| extra over for | | | | | | | |
| elbow | 12 | nr | 2.16 | 31.32 | 6.00 | 5.60 | 42.92 |
| tee | 6 | nr | 1.32 | 19.14 | 4.20 | 3.50 | 26.84 |
| tap connector | 6 | nr | 1.08 | 15.66 | 21.66 | 5.60 | 42.92 |
| 19mm diameter MPVC-U overflow pipe | 1 | m | 0.20 | 2.90 | 1.72 | 0.69 | 5.31 |
| extra over for | | | | | | | |
| elbow | 2 | nr | 0.36 | 5.22 | 2.52 | 1.16 | 8.90 |
| straight connector | 2 | nr | 0.40 | 5.80 | 3.74 | 1.43 | 10.97 |
| 32mm diameter polypropylene waste pipe | 1.50 | m | 0.30 | 4.35 | 1.63 | 0.90 | 6.88 |
| extra over for | | | | | | | |
| bend | 2 | nr | 0.36 | 5.22 | 1.84 | 1.06 | 8.12 |
| **Carried forward** | | | 15.34 | 222.43 | 83.59 | 45.90 | 351.92 |

| | Qty | Unit | Hours | Hours £ | Mat'ls £ | O & P £ | Total £ |
|---|---|---|---|---|---|---|---|
| **Brought forward** | | | 15.34 | 222.43 | 83.59 | 45.90 | 351.92 |
| **Traps** | | | | | | | |
| 32mm diameter polypropylene P trap | 3 | nr | 0.90 | 13.05 | 12.60 | 3.85 | 29.50 |
| **Sanitary fittings** | | | | | | | |
| Acrylic reinforced bath 1700mm long complete with chromium-plated grip handles, 40mm waste fitting, overflow, chain, plug bath panels, 20mm chromium-plated taps | 1 | nr | 2.75 | 39.88 | 121.74 | 24.24 | 185.86 |
| Vitreous china wash basin size 560 × 430mm, complete with 32mm waste fitting, overflow, chain, stay and plug, pair 13mm chromium-plated easy clean pillar taps, cast iron cantilever brackets and pedestal | 1 | nr | 2.20 | 31.90 | 131.51 | 24.51 | 187.92 |
| Shower cubicle size 788 × 842 × 2115mm high, anodised aluminium frame and safety glass, complete with 8.15kW instant electric shower with adjustable spray handset, slide bar and hose, soap dish and chromium-plated waste fittings | 1 | nr | 3.80 | 55.10 | 689.48 | 111.69 | 856.27 |
| **Carried forward** | | | 24.99 | 362.36 | 1,038.92 | 210.19 | 1,611.46 |

| | Qty | Unit | Hours | Hours £ | Mat'ls £ | O & P £ | Total £ |
|---|---|---|---|---|---|---|---|
| **Brought forward** | | | 24.99 | 362.36 | 1,038.92 | 210.19 | 1,611.46 |
| Bidet, free-standing with plain rim and chromium plated monobloc spray and waste fittings complete | 1 | nr | 2.50 | 36.25 | 258.79 | 44.26 | 339.30 |
| **Total** | | | 27.49 | 398.61 | 1,297.71 | 254.45 | 1,950.76 |

| | Qty | Unit | Hours | Hours £ | Mat'ls £ | O & P £ | Total £ |
|---|---|---|---|---|---|---|---|
| **Bathroom ADE with 1 bath (A), 1 lavatory basin (D) and 1 shower cubicle (E)** | | | | | | | |
| **Stripping out** | | | | | | | |
| Take out sanitary fittings from existing bathroom, including supply pipes, overflows and wastes, remove debris and make good to floors and walls to receive new fittings and pipework | | Item | 6.00 | 87.00 | 15.00 | 15.30 | 117.30 |
| **Pipework** | | | | | | | |
| 15mm diameter copper supply pipe | 15.80 | m | 3.16 | 45.82 | 27.02 | 10.93 | 83.77 |
| extra over for | | | | | | | |
| elbow | 10 | nr | 1.80 | 26.10 | 5.00 | 4.67 | 35.77 |
| tee | 4 | nr | 0.88 | 12.76 | 2.80 | 2.33 | 17.89 |
| tap connector | 6 | nr | 1.08 | 15.66 | 21.66 | 5.60 | 42.92 |
| **Carried forward** | | | 12.92 | 187.34 | 71.48 | 38.82 | 297.64 |

| | Qty | Unit | Hours | Hours £ | Mat'ls £ | O & P £ | Total £ |
|---|---|---|---|---|---|---|---|
| **Brought forward** | | | 12.92 | 187.34 | 71.48 | 38.82 | 297.64 |
| 19mm diameter MPVC-U overflow pipe | 0.50 | m | 0.10 | 1.45 | 0.86 | 0.35 | 2.66 |
| extra over for | | | | | | | |
| elbow | 1 | nr | 0.18 | 2.61 | 1.26 | 0.58 | 4.45 |
| straight connector | 1 | nr | 0.20 | 2.90 | 1.87 | 0.72 | 5.49 |
| 32mm diameter polypropylene waste pipe | 1.00 | m | 0.20 | 2.90 | 1.09 | 0.60 | 4.59 |
| extra over for | | | | | | | |
| bend | 2 | nr | 0.36 | 5.22 | 1.84 | 1.06 | 8.12 |
| **Traps** | | | | | | | |
| 32mm diameter polypropylene P trap | 3 | nr | 0.90 | 13.05 | 12.60 | 3.85 | 29.50 |
| **Sanitary fittings** | | | | | | | |
| Acrylic reinforced bath 1700mm long complete with chromium-plated grip handles, 40mm waste fitting, overflow, chain, plug bath panels, 20mm chromium-plated taps | 1 | nr | 2.75 | 39.88 | 171.74 | 31.74 | 243.36 |
| Vitreous china wash basin size 560 × 430mm, complete with 32mm waste fitting, overflow,chain, stay and plug, pair 13mm chromium-plated easy clean pillar taps, cast iron cantilever brackets and pedestal | 1 | nr | 2.20 | 31.90 | 131.51 | 24.51 | 187.92 |
| **Carried forward** | | | 19.81 | 287.25 | 394.25 | 102.22 | 783.72 |

| | Qty | Unit | Hours | Hours £ | Mat'ls £ | O & P £ | Total £ |
|---|---|---|---|---|---|---|---|
| **Brought forward** | | | 19.81 | 287.25 | 394.25 | 102.22 | 783.72 |
| Shower cubicle size 788 × 842 × 2115mm high, anodised aluminium frame and safety glass, complete with 8.15kW instant electric shower with adjustable spray handset, slide bar and hose, soap dish and chromium-plated waste fittings | 1 | nr | 3.80 | 55.10 | 689.48 | 111.69 | 856.27 |
| **Total** | | | 23.61 | 342.35 | 1,083.73 | 213.91 | 1,639.99 |

## SUMMARY OF BATHROOM PROJECT COSTS

| | Hours | Hours £ | Mat'ls £ | O & P £ | Total £ |
|---|---|---|---|---|---|
| Bath with hand shower, WC and lavatory basin | 19.94 | 289.13 | 650.73 | 140.98 | 1,080.84 |
| Bath with hand shower, WC, lavatory basin and bidet | 24.20 | 350.90 | 923.42 | 191.15 | 1,465.47 |
| Bath with hand shower, lavatory basin and bidet | 21.93 | 317.99 | 725.50 | 156.52 | 1,200.01 |
| Bath with hand shower and lavatory basin | 18.23 | 264.34 | 456.59 | 108.14 | 829.06 |
| Bath with hand shower, WC, lavatory basin and shower cubicle | 26.70 | 387.15 | 1,361.29 | 262.27 | 2,010.71 |
| Bath with hand shower, WC, lavatory basin, shower cubicle and bidet | 30.98 | 449.22 | 1,642.18 | 313.71 | 2,405.11 |
| Bath with hand shower, lavatory basin, shower cubicle and bidet | 27.49 | 398.61 | 1,428.92 | 274.13 | 2,101.65 |
| Bath with hand shower, lavatory basin and shower cubicle | 23.61 | 342.35 | 1,162.20 | 225.68 | 1,730.23 |
| Bath, WC and lavatory basin | 20.70 | 300.15 | 579.19 | 131.90 | 1,011.24 |
| Bath, WC and lavatory basin and bidet | 24.20 | 350.90 | 920.46 | 190.70 | 1,462.06 |
| Bath, lavatory basin and bidet | 21.97 | 318.57 | 593.98 | 136.88 | 1,049.43 |
| Bath and lavatory basin | 18.23 | 264.34 | 376.62 | 96.14 | 737.10 |

| | Qty | Unit | Hours | Hours £ | Mat'ls £ | O & P £ | Total £ |
|---|---|---|---|---|---|---|---|
| Bath, WC, lavatory basin and shower cubicle | | | 26.30 | 376.13 | 1,281.32 | 248.62 | 1,906.07 |
| Bath, WC, lavatory basin shower cubicle and bidet | | | 31.34 | 454.43 | 1,562.21 | 302.50 | 2,319.14 |
| Bath, lavatory basin, shower cubicle and bidet | | | 27.49 | 398.61 | 1,297.71 | 254.45 | 1,950.76 |
| Bath, lavatory basin and shower cubicle | | | 23.61 | 342.35 | 1,083.73 | 213.91 | 1,639.99 |

| | Qty | Unit | Hours | Hours £ | Mat'ls £ | O & P £ | Total £ |
|---|---|---|---|---|---|---|---|
| **EXTERNAL WASTE SYSTEMS** | | | | | | | |
| **MuPVC WASTE PIPES** | | | | | | | |
| **External waste system with solvent-welded joints for one storey-house** | | | | | | | |
| Take down existing external waste pipes, remove and prepare surfaces to receive new | | Item | 2.00 | 29.00 | 1.00 | 4.50 | 34.50 |
| 32mm diameter waste pipe, solvent-welded joints, clips at 500mm maximum centres, plugged to brickwork | 1.00 | m | 0.25 | 3.63 | 3.02 | 1.00 | 7.64 |
| extra over for bend | 1 | nr | 0.24 | 3.48 | 1.84 | 0.80 | 6.12 |
| Connection to back inlet, caulking bush | 1 | nr | 0.15 | 2.18 | 3.50 | 0.85 | 6.53 |
| 40mm diameter waste pipe, solvent-welded joints, clips at 500mm maximum centres, plugged to brickwork | 2.00 | m | 0.56 | 8.12 | 6.50 | 2.19 | 16.81 |
| extra over for bend | 2 | nr | 0.52 | 7.54 | 4.22 | 1.76 | 13.52 |
| Connection to back inlet, caulking bush | 2 | nr | 0.30 | 4.35 | 7.00 | 1.70 | 13.05 |
| **Total** | | | 4.02 | 58.29 | 27.08 | 12.81 | 98.18 |

| | Qty | Unit | Hours | Hours £ | Mat'ls £ | O & P £ | Total £ |
|---|---|---|---|---|---|---|---|
| **External waste system with solvent-welded joints for two-storey house** | | | | | | | |
| Take down existing external waste pipes, remove and prepare surfaces to receive new | | Item | 2.50 | 36.25 | 1.00 | 5.59 | 42.84 |
| 32mm diameter waste pipe, solvent-welded joints, clips at 500mm maximum centres, plugged to brickwork | 1.00 | m | 0.25 | 3.63 | 3.02 | 1.00 | 7.64 |
| extra over for bend | 1 | nr | 0.24 | 3.48 | 1.84 | 0.80 | 6.12 |
| Connection to back inlet, caulking bush | 1 | nr | 0.15 | 2.18 | 3.50 | 0.85 | 6.53 |
| 40mm diameter waste pipe, solvent-welded joints, clips at 500mm maximum centres, plugged to brickwork | 2.00 | m | 0.50 | 7.25 | 6.50 | 2.06 | 15.81 |
| extra over for bend | 2 | nr | 0.52 | 7.54 | 4.22 | 1.76 | 13.52 |
| Connection to back inlet, caulking bush | 2 | nr | 0.30 | 4.35 | 7.00 | 1.70 | 13.05 |
| **Total** | | | 4.46 | 64.67 | 27.08 | 13.76 | 105.51 |

| | Qty | Unit | Hours | Hours £ | Mat'ls £ | O & P £ | Total £ |
|---|---|---|---|---|---|---|---|
| **External waste system with solvent-welded joints for three-storey house** | | | | | | | |
| Take down existing external waste pipes, remove and prepare surfaces to receive new | | Item | 3.00 | 43.50 | 1.00 | 6.68 | 51.18 |
| 32mm diameter waste pipe, solvent-welded joints, clips at 500mm maximum centres, plugged to brickwork | 2.00 | m | 0.50 | 7.25 | 6.04 | 1.99 | 15.28 |
| extra over for bend | 2 | nr | 0.48 | 6.96 | 3.68 | 1.60 | 12.24 |
| Connection to back inlet, caulking bush | 2 | nr | 0.30 | 4.35 | 7.00 | 1.70 | 13.05 |
| 40mm diameter waste pipe, solvent-welded joints, clips at 500mm maximum centres, plugged to brickwork | 3.00 | m | 0.84 | 12.18 | 9.75 | 3.29 | 25.22 |
| extra over for bend | 3 | nr | 0.78 | 11.31 | 6.33 | 2.65 | 20.29 |
| Connection to back inlet, caulking bush | 3 | nr | 0.45 | 6.53 | 10.50 | 2.55 | 19.58 |
| **Total** | | | 6.35 | 92.08 | 44.30 | 20.46 | 156.83 |

| | Qty | Unit | Hours | Hours £ | Mat'ls £ | O & P £ | Total £ |
|---|---|---|---|---|---|---|---|
| **External waste system with push-fit joints for one-storey house** | | | | | | | |
| Take down existing external waste pipes, remove and prepare surfaces to receive new | | Item | 2.00 | 29.00 | 1.00 | 4.50 | 34.50 |
| 32mm diameter waste pipe, push-fit joints, clips at 500mm maximum centres, plugged to brickwork | 1.00 | m | 0.20 | 2.90 | 1.09 | 0.60 | 4.59 |
| extra over for bend | 1 | nr | 0.18 | 2.61 | 0.92 | 0.53 | 4.06 |
| Connection to back inlet, caulking bush | 1 | nr | 0.15 | 2.18 | 3.50 | 0.85 | 6.53 |
| 40mm diameter waste pipe, push-fit joints, clips at 500mm maximum centres, plugged to brickwork | 2.00 | m | 0.48 | 6.96 | 3.68 | 1.60 | 12.24 |
| extra over for bend | 2 | nr | 0.40 | 5.80 | 1.96 | 1.16 | 8.92 |
| Connection to back inlet, caulking bush | 2 | nr | 0.30 | 4.35 | 7.00 | 1.70 | 13.05 |
| **Total** | | | 3.71 | 53.80 | 19.15 | 10.94 | 83.89 |

| | Qty | Unit | Hours | Hours £ | Mat'ls £ | O & P £ | Total £ |
|---|---|---|---|---|---|---|---|
| **External waste system with push-fit joints for two-storey house** | | | | | | | |
| Take down existing external waste pipes, remove and prepare surfaces to receive new | | Item | 2.50 | 36.25 | 1.00 | 5.59 | 42.84 |
| 32mm diameter waste pipe, push-fit joints, clips at 500mm maximum centres, plugged to brickwork | 1.00 | m | 0.20 | 2.90 | 1.09 | 0.60 | 4.59 |
| extra over for bend | 1 | nr | 0.18 | 2.61 | 0.92 | 0.53 | 4.06 |
| Connection to back inlet, caulking bush | 1 | nr | 0.15 | 2.18 | 3.50 | 0.85 | 6.53 |
| 40mm diameter waste pipe, push-fit joints, clips at 500mm maximum centres, plugged to brickwork | 2.00 | m | 0.48 | 6.96 | 3.68 | 1.60 | 12.24 |
| extra over for bend | 2 | nr | 0.40 | 5.80 | 1.96 | 1.16 | 8.92 |
| Connection to back inlet, caulking bush | 2 | nr | 0.30 | 4.35 | 7.00 | 1.70 | 13.05 |
| **Total** | | | 4.21 | 61.05 | 19.15 | 12.03 | 92.22 |

| | Qty | Unit | Hours | Hours £ | Mat'ls £ | O & P £ | Total £ |
|---|---|---|---|---|---|---|---|
| **External waste system with push-fit joints for three-storey house** | | | | | | | |
| Take down existing external waste pipes, remove and prepare surfaces to receive new | | Item | 3.00 | 43.50 | 1.00 | 6.68 | 51.18 |
| 32mm diameter waste pipe, push-fit joints, clips at 500mm maximum centres, plugged to brickwork | 2.00 | m | 0.40 | 5.80 | 2.18 | 1.20 | 9.18 |
| extra over for bend | 2 | nr | 0.36 | 5.22 | 1.84 | 1.06 | 8.12 |
| Connection to back inlet, caulking bush | 2 | nr | 0.30 | 4.35 | 7.00 | 1.70 | 13.05 |
| 40mm diameter waste pipe, push-fit joints, clips at 500mm maximum centres, plugged to brickwork | 3.00 | m | 0.72 | 10.44 | 5.52 | 2.39 | 18.35 |
| extra over for bend | 3 | nr | 0.60 | 8.70 | 2.94 | 1.75 | 13.39 |
| Connection to back inlet, caulking bush | 3 | nr | 0.45 | 6.53 | 10.50 | 2.55 | 19.58 |
| **Total** | | | 5.83 | 84.54 | 30.98 | 17.33 | 132.84 |

| | Qty | Unit | Hours | Hours £ | Mat'ls £ | O & P £ | Total £ |
|---|---|---|---|---|---|---|---|
| **PVC-U SOIL PIPES** | | | | | | | |
| **External soil system with solvent-welded joints for one-storey house** | | | | | | | |
| Take down existing soil pipes, remove and prepare surfaces to receive new | | Item | 4.00 | 58.00 | 1.00 | 8.85 | 67.85 |
| 110mm diameter soil pipe, solvent-welded joints, holder-bats at 500mm maximum centres, plugged | 3.30 | m | 1.25 | 18.13 | 38.05 | 8.43 | 64.60 |
| extra over for offset | 1 | nr | 0.40 | 5.80 | 11.18 | 2.55 | 19.53 |
| Connection to vitrified clayware drain | 1 | nr | 0.40 | 5.80 | 17.32 | 3.47 | 26.59 |
| **Total** | | | 6.05 | 87.73 | 67.55 | 23.29 | 178.57 |

| | Qty | Unit | Hours | Hours £ | Mat'ls £ | O & P £ | Total £ |
|---|---|---|---|---|---|---|---|
| **External soil system with solvent-welded joints for two-storey house** | | | | | | | |
| Take down existing soil pipes, remove and prepare surfaces to receive new | | Item | 4.00 | 58.00 | 1.00 | 8.85 | 67.85 |
| 110mm diameter soil pipe, solvent-welded joints, holder-bats at 500mm maximum centres, plugged | 6.60 | m | 2.50 | 36.25 | 76.10 | 16.85 | 129.20 |
| **Carried forward** | | | 6.50 | 94.25 | 77.10 | 25.70 | 197.05 |

| | Qty | Unit | Hours | Hours £ | Mat'ls £ | O & P £ | Total £ |
|---|---|---|---|---|---|---|---|
| **Brought forward** | | | 6.50 | 94.25 | 77.10 | 25.70 | 197.05 |
| extra over for | | | | | | | |
| offset | 1 | nr | 0.40 | 5.80 | 11.18 | 2.55 | 19.53 |
| branch | 1 | nr | 0.40 | 5.80 | 19.31 | 3.77 | 28.88 |
| Connection to vitrified clayware drain | 1 | nr | 0.40 | 5.80 | 17.32 | 3.47 | 26.59 |
| **Total** | | | 7.70 | 111.65 | 124.91 | 35.48 | 272.04 |

| | Qty | Unit | Hours | Hours £ | Mat'ls £ | O & P £ | Total £ |
|---|---|---|---|---|---|---|---|
| **External soil system with solvent-welded joints for three-storey house** | | | | | | | |
| Take down existing soil pipes, remove and prepare surfaces to receive new | | Item | 6.00 | 87.00 | 1.00 | 13.20 | 101.20 |
| 110mm diameter soil pipe, solvent-welded joints, holder-bats at 500mm maximum centres, plugged | 9.30 | m | 3.53 | 51.19 | 107.23 | 23.76 | 182.18 |
| extra over for | | | | | | | |
| offset | 1 | nr | 0.40 | 5.80 | 11.18 | 2.55 | 19.53 |
| branch | 2 | nr | 0.80 | 11.60 | 38.62 | 7.53 | 57.75 |
| Connection to vitrified clayware drain | 1 | nr | 0.40 | 5.80 | 17.32 | 3.47 | 26.59 |
| **Total** | | | 11.13 | 161.39 | 175.35 | 50.51 | 387.25 |

| | Qty | Unit | Hours | Hours £ | Mat'ls £ | O & P £ | Total £ |
|---|---|---|---|---|---|---|---|
| **CAST IRON SOIL PIPES** | | | | | | | |
| **External soil system with flexible joints for one-storey house** | | | | | | | |
| Take down existing soil pipes, remove and prepare surfaces to receive new | | Item | 4.00 | 58.00 | 1.00 | 8.85 | 67.85 |
| 100mm diameter soil pipe, flexible joints, holder-bats at 500mm maximum centres, plugged | 3.30 | m | 2.15 | 31.18 | 97.25 | 19.26 | 147.69 |
| extra over for offset | 1 | nr | 0.55 | 7.98 | 17.08 | 3.76 | 28.81 |
| Connection to vitrified clayware drain | 1 | nr | 0.60 | 8.70 | 12.39 | 3.16 | 24.25 |
| **Total** | | | 7.30 | 105.85 | 127.72 | 35.04 | 268.61 |

| | Qty | Unit | Hours | Hours £ | Mat'ls £ | O & P £ | Total £ |
|---|---|---|---|---|---|---|---|
| **External soil system with flexible joints for two-storey house** | | | | | | | |
| Take down existing soil pipes, remove and prepare surfaces to receive new | | Item | 4.00 | 58.00 | 1.00 | 8.85 | 67.85 |
| 100mm diameter soil pipe, flexible joints, holder-bats at 500mm maximum centres, plugged | 6.60 | m | 4.29 | 62.21 | 194.50 | 38.51 | 295.21 |
| **Carried forward** | | | 8.29 | 120.21 | 195.50 | 47.36 | 363.06 |

| | Qty | Unit | Hours | Hours £ | Mat'ls £ | O & P £ | Total £ |
|---|---|---|---|---|---|---|---|
| **Brought forward** | | | 8.29 | 120.21 | 195.50 | 47.36 | 363.06 |
| extra over for | | | | | | | |
| offset | 1 | nr | 0.55 | 7.98 | 17.08 | 3.76 | 28.81 |
| branch | 1 | nr | 0.60 | 8.70 | 21.05 | 3.87 | 29.65 |
| Connection to vitrified clayware drain | 1 | nr | 0.60 | 8.70 | 12.39 | 3.16 | 24.25 |
| **Total** | | | 10.04 | 145.59 | 246.02 | 58.15 | 445.77 |

| | Qty | Unit | Hours | Hours £ | Mat'ls £ | O & P £ | Total £ |
|---|---|---|---|---|---|---|---|
| **External soil system with flexible joints for three-storey house** | | | | | | | |
| Take down existing soil pipes, remove and prepare surfaces to receive new | | Item | 6.00 | 87.00 | 1.00 | 13.20 | 101.20 |
| 100mm diameter soil pipe, flexible joints, holder-bats at 500mm maximum centres, plugged | 9.90 | m | 6.05 | 87.73 | 274.07 | 54.27 | 416.06 |
| extra over for | | | | | | | |
| offset | 1 | nr | 0.55 | 7.98 | 17.08 | 3.76 | 28.81 |
| branch | 2 | nr | 1.20 | 17.40 | 42.10 | 8.93 | 68.43 |
| Connection to vitrified clayware drain | 1 | nr | 0.60 | 8.70 | 12.39 | 3.16 | 24.25 |
| **Total** | | | 14.40 | 208.80 | 346.64 | 83.32 | 638.76 |

## SUMMARY OF EXTERNAL WASTE PROJECT COSTS

| | Hours | Hours £ | Mat'ls £ | O & P £ | Total £ |
|---|---|---|---|---|---|
| **MuPVC waste system with solvent-welded joints** | | | | | |
| One storey house | 4.02 | 58.29 | 27.08 | 12.81 | 98.18 |
| Two storey house | 4.46 | 64.67 | 27.08 | 13.76 | 105.51 |
| Three storey house | 6.35 | 92.08 | 44.30 | 20.46 | 156.83 |
| **MuPVC waste system with push-fit joints** | | | | | |
| One storey house | 3.71 | 53.80 | 19.15 | 10.94 | 83.89 |
| Two storey house | 4.21 | 61.05 | 19.15 | 12.03 | 92.22 |
| Three storey house | 5.83 | 84.54 | 30.98 | 17.33 | 132.84 |
| **PVC-U soil system with solvent-welded joints** | | | | | |
| One storey house | 6.05 | 87.73 | 67.55 | 23.29 | 178.57 |
| Two storey house | 7.70 | 111.65 | 124.91 | 35.48 | 272.04 |
| Three storey house | 11.13 | 161.39 | 175.35 | 50.51 | 387.25 |
| **Cast iron soil system with flexible joints** | | | | | |
| One storey house | 7.30 | 105.85 | 127.72 | 35.04 | 268.61 |
| Two storey house | 10.04 | 145.58 | 242.05 | 58.14 | 445.77 |
| Three storey house | 14.40 | 208.80 | 346.64 | 83.32 | 638.76 |

| | Qty | Unit | Hours | Hours £ | Mat'ls £ | O & P £ | Total £ |
|---|---|---|---|---|---|---|---|
| **CENTRAL HEATING SYSTEMS** | | | | | | | |
| **Central heating system for one-storey house, overall size 7 × 6m** | | | | | | | |
| 15mm diameter copper pipe, capillary joints and fittings, clips at 1250mm maximum centres | 66.00 | m | 13.20 | 191.40 | 112.86 | 45.64 | 349.90 |
| extra over for | | | | | | | |
| elbow | 18 | nr | 3.24 | 46.98 | 9.00 | 8.40 | 64.38 |
| tee | 13 | nr | 2.86 | 41.47 | 9.10 | 7.59 | 58.16 |
| Connection to central heating boiler, 15mm diameter pipe | 4 | nr | 1.12 | 16.24 | 12.00 | 4.24 | 32.48 |
| Connection to storage tank, 15mm diameter pipe | 2 | nr | 0.56 | 8.12 | 10.00 | 2.72 | 20.84 |
| Break into existing copper pipe and insert 15mm tee | 1 | nr | 0.90 | 13.05 | 2.60 | 2.35 | 18.00 |
| Gas-fired wall-mounted central combination boiler, 30,400 to 81,900 Btu, fan flue | 1 | nr | 4.80 | 69.60 | 651.68 | 108.19 | 829.47 |
| Pressed steel radiator fixed to brickwork with concealed brackets, size | | | | | | | |
| single, 600 × 1400mm | 1 | nr | 1.35 | 19.58 | 41.64 | 9.18 | 70.40 |
| double, 600 × 1000mm | 2 | nr | 2.50 | 36.25 | 106.38 | 21.39 | 164.02 |
| double, 600 × 1800mm | 2 | nr | 3.00 | 43.50 | 266.56 | 46.51 | 356.57 |
| Chromium-plated radiator valve, 15mm × 1/2in | 5 | nr | 1.50 | 21.75 | 92.20 | 17.09 | 131.04 |
| **Carried forward** | | | 35.03 | 507.94 | 1,314.02 | 273.29 | 2,095.25 |

| | Qty | Unit | Hours | Hours £ | Mat'ls £ | O & P £ | Total £ |
|---|---|---|---|---|---|---|---|
| Brought forward | | | 35.03 | 507.94 | 1,314.02 | 273.29 | 2,095.25 |
| Lockshield radiator valve, 15mm × 1/2in | 5 | nr | 1.50 | 21.75 | 31.20 | 7.94 | 60.89 |
| Galvanised steel tank ref. T25/1, 36 litres | 1 | nr | 0.65 | 9.43 | 62.02 | 10.72 | 82.16 |
| Hole through existing 100mm thick plastered block wall for two small pipes and make good | 1 | nr | 0.40 | 5.80 | 0.20 | 0.90 | 6.90 |
| Hole through existing stud partition plasterboard and skim both sides for two small pipes and make good | 3 | nr | 0.60 | 8.70 | 0.30 | 1.35 | 10.35 |
| Hole through existing plasterboard and skim ceiling for two small pipes and make good | 1 | nr | 0.20 | 2.90 | 0.10 | 0.45 | 3.45 |
| **Total** | | | 38.38 | 556.52 | 1,407.84 | 294.65 | 2,259.00 |

| | Qty | Unit | Hours | Hours £ | Mat'ls £ | O & P £ | Total £ |
|---|---|---|---|---|---|---|---|
| **Central heating system for one-storey house, overall size 8 × 7m** | | | | | | | |
| 15mm diameter copper pipe, capillary joints and fittings, clips at 1250mm maximum centres | 72.00 | m | 14.40 | 208.80 | 123.12 | 49.79 | 381.71 |
| **Carried forward** | | | 15.84 | 173.29 | 135.36 | 46.30 | 354.95 |

| | Qty | Unit | Hours | Hours £ | Mat'ls £ | O & P £ | Total £ |
|---|---|---|---|---|---|---|---|
| **Brought forward** | | | 15.84 | 173.29 | 135.36 | 46.30 | 354.95 |
| extra over for | | | | | | | |
| elbow | 18 | nr | 3.24 | 46.98 | 9.00 | 8.40 | 64.38 |
| tee | 13 | nr | 2.86 | 41.47 | 9.10 | 7.59 | 58.16 |
| Connection to central heating boiler, 15mm diameter pipe | 4 | nr | 1.12 | 16.24 | 12.00 | 4.24 | 32.48 |
| Connection to storage tank, 15mm diameter pipe | 2 | nr | 0.56 | 8.12 | 10.00 | 2.72 | 20.84 |
| Break into existing copper pipe and insert 15mm tee | 1 | nr | 0.90 | 13.05 | 2.60 | 2.35 | 18.00 |
| Gas-fired wall-mounted central combination boiler, 30,400 to 81,900 Btu, fan flue | 1 | nr | 4.80 | 69.60 | 651.68 | 108.19 | 829.47 |
| Pressed steel radiator fixed to brickwork with concealed brackets, size | | | | | | | |
| single, 600 × 1400mm | 1 | nr | 1.35 | 19.58 | 41.64 | 9.18 | 70.40 |
| double, 600 × 1000mm | 3 | nr | 3.75 | 54.38 | 159.57 | 32.09 | 246.04 |
| double, 600 × 1800mm | 2 | nr | 3.00 | 43.50 | 266.56 | 46.51 | 356.57 |
| Chromium-plated radiator valve, 15mm × 1/2in | 6 | nr | 1.80 | 26.10 | 110.64 | 20.51 | 157.25 |
| Lockshield radiator valve, 15mm × 1/2in | 6 | nr | 1.80 | 26.10 | 37.44 | 9.53 | 73.07 |
| Galvanised steel tank ref. T25/1, 36 litres | 1 | nr | 0.65 | 9.43 | 62.02 | 10.72 | 82.16 |
| **Carried forward** | | | 41.67 | 547.83 | 1,507.61 | 308.32 | 2,363.75 |

| | Qty | Unit | Hours | Hours £ | Mat'ls £ | O & P £ | Total £ |
|---|---|---|---|---|---|---|---|
| **Brought forward** | | | 41.67 | 547.83 | 1,507.61 | 308.32 | 2,363.75 |
| Hole through existing 100mm thick plastered block wall for two pipes and make good | 1 | nr | 0.40 | 5.80 | 0.20 | 0.90 | 6.90 |
| Hole through existing stud partition plasterboard and skim both sides for two small pipes and make good | 3 | nr | 0.60 | 8.70 | 0.30 | 1.35 | 10.35 |
| Hole through existing plasterboard and skim ceiling for two small pipes and make good | 1 | nr | 0.20 | 2.90 | 0.10 | 0.45 | 3.45 |
| **Total** | | | 42.87 | 565.23 | 1,508.21 | 311.02 | 2,384.45 |

| | Qty | Unit | Hours | Hours £ | Mat'ls £ | O & P £ | Total £ |
|---|---|---|---|---|---|---|---|
| **Central heating system for one-storey house, overall size 9 × 7m** | | | | | | | |
| 15mm diameter copper pipe, capillary joints and fittings, clips at 1250mm maximum centres | 77.00 | m | 15.40 | 223.30 | 131.67 | 53.25 | 408.22 |
| extra over for | | | | | | | |
| elbow | 18 | nr | 3.24 | 46.98 | 9.00 | 8.40 | 64.38 |
| tee | 13 | nr | 2.86 | 41.47 | 9.10 | 7.59 | 58.16 |
| Connection to central heating boiler, 15mm diameter pipe | 4 | nr | 1.12 | 16.24 | 12.00 | 4.24 | 32.48 |
| **Carried forward** | | | 22.62 | 327.99 | 161.77 | 73.46 | 563.22 |

| | Qty | Unit | Hours | Hours £ | Mat'ls £ | O & P £ | Total £ |
|---|---|---|---|---|---|---|---|
| **Brought forward** | | | 22.62 | 327.99 | 161.77 | 73.46 | 563.22 |
| Connection to storage tank, 15mm diameter pipe | 2 | nr | 0.56 | 8.12 | 10.00 | 2.72 | 20.84 |
| Break into existing copper pipe and insert 15mm tee | 1 | nr | 0.90 | 13.05 | 2.60 | 2.35 | 18.00 |
| Gas-fired wall-mounted central combination boiler, 35,500-95,000 Btu, condensing flue | 1 | nr | 4.80 | 69.60 | 696.68 | 114.94 | 881.22 |
| Pressed steel radiator fixed to brickwork with concealed brackets, size | | | | | | | |
| single, 600 × 1400mm | 1 | nr | 1.35 | 19.58 | 41.64 | 9.18 | 70.40 |
| double, 600 × 1000mm | 4 | nr | 5.00 | 72.50 | 212.76 | 42.79 | 328.05 |
| double, 600 × 1800mm | 2 | nr | 3.00 | 43.50 | 266.56 | 46.51 | 356.57 |
| Chromium-plated radiator valve, 15mm × 1/2in | 7 | nr | 2.10 | 30.45 | 129.08 | 23.93 | 183.46 |
| Lockshield radiator valve, 15mm × 1/2in | 7 | nr | 2.10 | 30.45 | 43.68 | 11.12 | 85.25 |
| Galvanised steel tank ref. T25/1, 36 litres | 1 | nr | 0.65 | 9.43 | 62.02 | 10.72 | 82.16 |
| Hole through existing 100mm thick plastered block wall for two small pipes and make good | 1 | nr | 0.40 | 5.80 | 0.20 | 0.90 | 6.90 |
| **Carried forward** | | | 43.48 | 560.86 | 930.31 | 223.67 | 1,714.84 |

| | Qty | Unit | Hours | Hours £ | Mat'ls £ | O & P £ | Total £ |
|---|---|---|---|---|---|---|---|
| **Brought forward** | | | 43.48 | 560.86 | 930.31 | 223.67 | 1,714.84 |
| Hole through existing stud partition plasterboard and skim both sides for two small pipes and make good | 3 | nr | 0.60 | 8.70 | 0.30 | 1.35 | 10.35 |
| Hole through existing plasterboard and skim ceiling for two small pipes and make good | 1 | nr | 0.20 | 2.90 | 0.10 | 0.45 | 3.45 |
| **Total** | | | 44.28 | 572.46 | 930.71 | 225.47 | 1,728.64 |

| | Qty | Unit | Hours | Hours £ | Mat'ls £ | O & P £ | Total £ |
|---|---|---|---|---|---|---|---|
| **Central heating system for two-storey house, overall size 7 × 7m** | | | | | | | |
| 15mm diameter copper pipe, capillary joints and fittings, clips at 1250mm maximum centres | 113.00 | m | 22.60 | 247.24 | 193.23 | 66.07 | 506.55 |
| extra over for | | | | | | | |
| elbow | 29 | nr | 5.22 | 75.69 | 14.50 | 13.53 | 103.72 |
| tee | 15 | nr | 3.30 | 47.85 | 10.50 | 8.75 | 67.10 |
| Connection to central heating boiler, 15mm diameter pipe | 4 | nr | 1.12 | 16.24 | 12.00 | 4.24 | 32.48 |
| Connection to storage tank, 15mm diameter pipe | 2 | nr | 0.56 | 8.12 | 10.00 | 2.72 | 20.84 |
| **Carried forward** | | | 32.80 | 395.14 | 240.23 | 95.31 | 730.68 |

| | Qty | Unit | Hours | Hours £ | Mat'ls £ | O & P £ | Total £ |
|---|---|---|---|---|---|---|---|
| **Brought forward** | | | 32.80 | 395.14 | 240.23 | 95.31 | 730.68 |
| Break into existing copper pipe and insert 15mm tee | 1 | nr | 0.90 | 13.05 | 2.60 | 2.35 | 18.00 |
| Gas-fired wall-mounted central combination boiler, 35,500-95,000 Btu, fan flue | 1 | nr | 4.80 | 69.60 | 696.68 | 114.94 | 881.22 |
| Pressed steel radiator fixed to brickwork with concealed brackets, size | | | | | | | |
| single, 600 × 1400mm | 1 | nr | 1.35 | 19.58 | 41.64 | 9.18 | 70.40 |
| double, 600 × 1000mm | 5 | nr | 6.25 | 90.63 | 265.95 | 53.49 | 410.06 |
| double, 600 × 1800mm | 2 | nr | 3.00 | 43.50 | 266.56 | 46.51 | 356.57 |
| Chromium-plated radiator valve, 15mm × 1/2in | 8 | nr | 2.40 | 34.80 | 65.60 | 15.06 | 115.46 |
| Lockshield radiator valve, 15mm × 1/2in | 8 | nr | 2.40 | 34.80 | 49.92 | 12.71 | 97.43 |
| Galvanised steel tank ref. T25/1, 86 litres | 1 | nr | 0.75 | 10.88 | 76.01 | 13.03 | 99.92 |
| Hole through existing 100mm thick plastered block wall for two small pipes and make good | 2 | nr | 0.80 | 11.60 | 0.40 | 1.80 | 13.80 |
| Hole through existing stud partition plasterboard and skim both sides for two small pipes and make good | 5 | nr | 1.00 | 14.50 | 0.50 | 2.25 | 17.25 |
| **Carried forward** | | | 56.45 | 738.07 | 1,706.09 | 366.63 | 2,810.78 |

| | Qty | Unit | Hours | Hours £ | Mat'ls £ | O & P £ | Total £ |
|---|---|---|---|---|---|---|---|
| **Brought forward** | | | 56.45 | 738.07 | 1,706.09 | 366.63 | 2,810.78 |
| Hole through existing plasterboard and skim ceiling for two small pipes and make good | 2 | nr | 0.40 | 5.80 | 0.20 | 0.90 | 6.90 |
| Hole through existing softwood flooring for two small pipes and make good | 1 | nr | 0.20 | 2.90 | 0.10 | 0.45 | 3.45 |
| **Total** | | | 57.05 | 746.77 | 1,706.39 | 367.98 | 2,821.13 |

| | Qty | Unit | Hours | Hours £ | Mat'ls £ | O & P £ | Total £ |
|---|---|---|---|---|---|---|---|
| **Central heating system for two-storey house, overall size 8 × 7m** | | | | | | | |
| 15mm diameter copper pipe, capillary joints and fittings, clips at 1250mm maximum centres | 116.00 | m | 23.20 | 336.40 | 198.36 | 80.21 | 614.97 |
| extra over for | | | | | | | |
| elbow | 29 | nr | 5.22 | 75.69 | 14.50 | 13.53 | 103.72 |
| tee | 15 | nr | 3.30 | 47.85 | 10.50 | 8.75 | 67.10 |
| Connection to central heating boiler, 15mm diameter pipe | 4 | nr | 1.12 | 16.24 | 12.00 | 4.24 | 32.48 |
| Connection to storage tank, 15mm diameter pipe | 2 | nr | 0.56 | 8.12 | 10.00 | 2.72 | 20.84 |
| Break into existing copper pipe and insert 15mm tee | 1 | nr | 0.90 | 13.05 | 2.60 | 2.35 | 18.00 |
| **Carried forward** | | | 34.30 | 497.35 | 247.96 | 111.80 | 857.11 |

| | Qty | Unit | Hours | Hours £ | Mat'ls £ | O & P £ | Total £ |
|---|---|---|---|---|---|---|---|
| **Brought forward** | | | 34.30 | 497.35 | 247.96 | 111.80 | 857.11 |
| Gas-fired wall-mounted central combination boiler, 22,500-92,800 Btu, condensing flue | 1 | nr | 4.80 | 69.60 | 815.68 | 132.79 | 1,018.07 |
| Pressed steel radiator fixed to brickwork with concealed brackets, size | | | | | | | |
| single, 600 × 1400mm | 1 | nr | 2.70 | 39.15 | 83.28 | 18.36 | 140.79 |
| double, 600 × 1000mm | 5 | nr | 6.25 | 90.63 | 265.95 | 53.49 | 410.06 |
| double, 600 × 1800mm | 2 | nr | 3.00 | 43.50 | 266.56 | 46.51 | 356.57 |
| Chromium-plated radiator valve, 15mm × 1/2in | 9 | nr | 2.70 | 39.15 | 165.96 | 30.77 | 235.88 |
| Lockshield radiator valve, 15mm × 1/2in | 9 | nr | 2.70 | 39.15 | 58.40 | 14.63 | 112.18 |
| Galvanised steel tank ref. T25/1, 86 litres | 1 | nr | 0.75 | 10.88 | 76.01 | 13.03 | 99.92 |
| Hole through existing 100mm thick plastered block wall for two small pipes and make good | 2 | nr | 0.80 | 11.60 | 0.40 | 1.80 | 13.80 |
| Hole through existing stud partition plasterboard and skim both sides for two small pipes and make good | 5 | nr | 1.00 | 14.50 | 0.50 | 2.25 | 17.25 |
| Hole through existing plasterboard and skim ceiling for two small pipes and make good | 2 | nr | 0.40 | 5.80 | 0.20 | 0.90 | 6.90 |
| **Carried forward** | | | 59.40 | 861.30 | 1,980.90 | 426.33 | 3,268.53 |

| | Qty | Unit | Hours | Hours £ | Mat'ls £ | O & P £ | Total £ |
|---|---|---|---|---|---|---|---|
| **Brought forward** | | | 59.40 | 861.30 | 1,980.90 | 426.33 | 3,268.53 |
| Hole through existing softwood flooring for two small pipes and make good | 1 | nr | 0.20 | 2.90 | 0.10 | 0.45 | 3.45 |
| **Total** | | | 59.60 | 864.20 | 1,981.00 | 426.78 | 3,271.98 |

| | Qty | Unit | Hours | Hours £ | Mat'ls £ | O & P £ | Total £ |
|---|---|---|---|---|---|---|---|
| **Central heating system for two-storey house, overall size 9 × 8m** | | | | | | | |
| 15mm diameter copper pipe, capillary joints and fittings, clips at 1250mm maximum centres | 127.00 | m | 25.40 | 368.30 | 217.17 | 87.82 | 673.29 |
| extra over for | | | | | | | |
| elbow | 29 | nr | 5.22 | 75.69 | 14.50 | 13.53 | 103.72 |
| tee | 15 | nr | 3.30 | 47.85 | 10.50 | 8.75 | 67.10 |
| Connection to central heating boiler, 15mm diameter pipe | 4 | nr | 1.12 | 16.24 | 12.00 | 4.24 | 32.48 |
| Connection to storage tank, 15mm diameter pipe | 2 | nr | 0.56 | 8.12 | 10.00 | 2.72 | 20.84 |
| Break into existing copper pipe and insert 15mm tee | 1 | nr | 0.90 | 13.05 | 2.60 | 2.35 | 18.00 |
| Gas-fired wall-mounted central combination boiler, 22,500-92,800 Btu, condensing flue | 1 | nr | 4.80 | 69.60 | 815.68 | 132.79 | 1,018.07 |
| **Carried forward** | | | 41.30 | 598.85 | 1,082.45 | 252.20 | 1,933.50 |

| | Qty | Unit | Hours | Hours £ | Mat'ls £ | O & P £ | Total £ |
|---|---|---|---|---|---|---|---|
| **Brought forward** | | | 41.30 | 598.85 | 1,082.45 | 252.20 | 1,933.50 |
| Pressed steel radiator fixed to brickwork with concealed brackets, size | | | | | | | |
| single, 600 × 1400mm | 2 | nr | 2.70 | 39.15 | 83.28 | 18.36 | 140.79 |
| double, 600 × 1000mm | 6 | nr | 5.00 | 72.50 | 319.14 | 58.75 | 450.39 |
| double, 600 × 1800mm | 2 | nr | 3.00 | 43.50 | 266.56 | 46.51 | 356.57 |
| Chromium-plated radiator valve, 15mm × 1/2in | 10 | nr | 3.00 | 43.50 | 184.40 | 34.19 | 262.09 |
| Lockshield radiator valve, 15mm × 1/2in | 10 | nr | 3.00 | 43.50 | 66.00 | 16.43 | 125.93 |
| Galvanised steel tank ref. T25/1, 86 litres | 1 | nr | 0.75 | 10.88 | 76.01 | 13.03 | 99.92 |
| Hole through existing 100mm thick plastered block wall for two small pipes and make good | 2 | nr | 0.80 | 11.60 | 0.40 | 1.80 | 13.80 |
| Hole through existing stud partition plasterboard and skim both sides for two small pipes and make good | 5 | nr | 1.00 | 14.50 | 0.50 | 2.25 | 17.25 |
| Hole through existing plasterboard and skim ceiling for two small pipes and make good | 2 | nr | 0.40 | 5.80 | 0.20 | 0.90 | 6.90 |
| **Carried forward** | | | 60.95 | 883.78 | 2,078.94 | 444.41 | 3,407.13 |

| | Qty | Unit | Hours | Hours £ | Mat'ls £ | O & P £ | Total £ |
|---|---|---|---|---|---|---|---|
| **Brought forward** | | | 60.95 | 883.78 | 2,078.94 | 444.41 | 3,407.13 |
| Hole through existing softwood flooring for two small pipes and make good | 1 | nr | 0.20 | 2.90 | 0.10 | 0.45 | 3.45 |
| **Total** | | | 61.15 | 886.68 | 2,079.04 | 444.86 | 3,410.58 |

| | Qty | Unit | Hours | Hours £ | Mat'ls £ | O & P £ | Total £ |
|---|---|---|---|---|---|---|---|
| **Central heating system for three-storey house, overall size 8 × 8m** | | | | | | | |
| 15mm diameter copper pipe, capillary joints and fittings, clips at 1250mm maximum centres | 202.00 | m | 40.40 | 585.80 | 345.42 | 139.68 | 1,070.90 |
| extra over for | | | | | | | |
| elbow | 41 | nr | 7.38 | 107.01 | 20.50 | 19.13 | 146.64 |
| tee | 25 | nr | 5.50 | 79.75 | 17.50 | 14.59 | 111.84 |
| Connection to central heating boiler, 15mm diameter pipe | 4 | nr | 1.12 | 16.24 | 12.00 | 4.24 | 32.48 |
| Connection to storage tank, 15mm diameter pipe | 2 | nr | 0.56 | 8.12 | 10.00 | 2.72 | 20.84 |
| Break into existing copper pipe and insert 15mm tee | 1 | nr | 0.90 | 13.05 | 2.60 | 2.35 | 18.00 |
| Gas-fired wall-mounted central combination boiler, 25,900-93,800 Btu, condensing flue | 1 | nr | 4.80 | 69.60 | 843.68 | 136.99 | 1,050.27 |
| **Carried forward** | | | 60.66 | 809.97 | 408.02 | 182.70 | 1,400.69 |

| | Qty | Unit | Hours | Hours £ | Mat'ls £ | O & P £ | Total £ |
|---|---|---|---|---|---|---|---|
| **Brought forward** | | | 60.66 | 809.97 | 408.02 | 182.70 | 1,400.69 |
| Pressed steel radiator fixed to brickwork with concealed brackets, size | | | | | | | |
| single, 600 × 1400mm | 2 | nr | 2.70 | 39.15 | 83.28 | 18.36 | 140.79 |
| double, 600 × 1000mm | 7 | nr | 8.75 | 126.88 | 372.33 | 74.88 | 574.09 |
| double, 600 × 1800mm | 3 | nr | 4.50 | 65.25 | 399.84 | 69.76 | 534.85 |
| Chromium-plated radiator valve, 15mm × 1/2in | 12 | nr | 3.60 | 52.20 | 221.28 | 41.02 | 314.50 |
| Lockshield radiator valve, 15mm × 1/2in | 12 | nr | 3.60 | 52.20 | 78.20 | 19.56 | 149.96 |
| Galvanised steel tank ref. T25/1, 86 litres | 1 | nr | 0.75 | 10.88 | 76.01 | 13.03 | 99.92 |
| Hole through existing 100mm thick plastered block wall for two small pipes and make good | 2 | nr | 0.80 | 11.60 | 0.40 | 1.80 | 13.80 |
| Hole through existing stud partition plasterboard and skim both sides for two small and make good | 9 | nr | 1.80 | 26.10 | 0.90 | 4.05 | 31.05 |
| Hole through existing plasterboard and skim ceiling for two small pipes and make good | 3 | nr | 0.60 | 8.70 | 0.30 | 1.35 | 10.35 |
| **Carried forward** | | | 87.76 | 1,202.92 | 1,640.56 | 426.52 | 3,270.00 |

| | Qty | Unit | Hours | Hours £ | Mat'ls £ | O & P £ | Total £ |
|---|---|---|---|---|---|---|---|
| **Brought forward** | | | 87.76 | 1,202.92 | 1,640.56 | 426.52 | 3,270.00 |
| Hole through existing softwood flooring for two small pipes and make good | 2 | nr | 0.40 | 5.80 | 0.20 | 0.90 | 6.90 |
| **Total** | | | 88.16 | 1,208.72 | 1,640.76 | 427.42 | 3,276.90 |

| | Qty | Unit | Hours | Hours £ | Mat'ls £ | O & P £ | Total £ |
|---|---|---|---|---|---|---|---|
| **Central heating system for three-storey house, overall size 9 × 8m** | | | | | | | |
| 15mm diameter copper pipe, capillary joints and fittings, clips at 1250mm maximum centres | 217.00 | m | 41.40 | 600.30 | 371.07 | 145.71 | 1,117.08 |
| extra over for | | | | | | | |
| elbow | 41 | nr | 7.38 | 107.01 | 20.50 | 19.13 | 146.64 |
| tee | 25 | nr | 5.50 | 79.75 | 17.50 | 14.59 | 111.84 |
| Connection to central heating boiler, 15mm diameter pipe | 4 | nr | 1.12 | 16.24 | 12.00 | 4.24 | 32.48 |
| Connection to storage tank, 15mm diameter pipe | 2 | nr | 0.56 | 8.12 | 10.00 | 2.72 | 20.84 |
| Break into existing copper pipe and insert 15mm tee | 1 | nr | 0.90 | 13.05 | 2.60 | 2.35 | 18.00 |
| Gas-fired wall-mounted central combination boiler, 25,900-93,800 Btu, condensing flue | 1 | nr | 4.80 | 69.60 | 843.68 | 136.99 | 1,050.27 |
| **Carried forward** | | | 61.66 | 894.07 | 1,277.35 | 325.71 | 2,497.13 |

| | Qty | Unit | Hours | Hours £ | Mat'ls £ | O & P £ | Total £ |
|---|---|---|---|---|---|---|---|
| **Brought forward** | | | 61.66 | 894.07 | 1,277.35 | 325.71 | 2,497.13 |
| Pressed steel radiator fixed to brickwork with concealed brackets, size | | | | | | | |
| single, 600 × 1400mm | 3 | nr | 4.05 | 58.73 | 124.92 | 27.55 | 211.19 |
| double, 600 × 1000mm | 8 | nr | 10.00 | 145.00 | 425.52 | 85.58 | 656.10 |
| double, 600 × 1800mm | 3 | nr | 4.50 | 65.25 | 399.84 | 69.76 | 534.85 |
| Chromium-plated radiator valve, 15mm × 1/2in | 14 | nr | 4.20 | 60.90 | 258.16 | 47.86 | 366.92 |
| Lockshield radiator valve, 15mm × 1/2in | 14 | nr | 4.20 | 60.90 | 92.40 | 23.00 | 176.30 |
| Galvanised steel tank ref. T25/1, 86 litres | 1 | nr | 0.75 | 10.88 | 76.01 | 13.03 | 99.92 |
| Hole through existing 100mm thick plastered block wall for two small pipes and make good | 2 | nr | 0.80 | 11.60 | 0.40 | 1.80 | 13.80 |
| Hole through existing stud partition plasterboard and skim both sides for two small and make good | 9 | nr | 1.80 | 26.10 | 0.90 | 4.05 | 31.05 |
| Hole through existing plasterboard and skim ceiling for two small pipes and make good | 3 | nr | 0.60 | 8.70 | 0.30 | 1.35 | 10.35 |
| **Carried forward** | | | 92.56 | 1,342.12 | 2,655.80 | 599.69 | 4,597.61 |

| | Qty | Unit | Hours | Hours £ | Mat'ls £ | O & P £ | Total £ |
|---|---|---|---|---|---|---|---|
| **Brought forward** | | | 92.56 | 1,342.12 | 2,655.80 | 599.69 | 4,597.61 |
| Hole through existing softwood flooring for two small pipes and make good | 2 | nr | 0.40 | 5.80 | 0.20 | 0.90 | 6.90 |
| **Total** | | | 92.96 | 1,347.92 | 2,656.00 | 600.59 | 4,604.51 |

| | Qty | Unit | Hours | Hours £ | Mat'ls £ | O & P £ | Total £ |
|---|---|---|---|---|---|---|---|
| **Central heating system for three-storey house, overall size 10 × 9m** | | | | | | | |
| 15mm diameter copper pipe, capillary joints and fittings, clips at 1250mm maximum centres | 224.00 | m | 44.80 | 649.60 | 383.04 | 154.90 | 1,187.54 |
| extra over for | | | | | | | |
| elbow | 41 | nr | 7.38 | 107.01 | 20.50 | 19.13 | 146.64 |
| tee | 25 | nr | 5.50 | 79.75 | 17.50 | 14.59 | 111.84 |
| Connection to central heating boiler, 15mm diameter pipe | 4 | nr | 1.12 | 16.24 | 12.00 | 4.24 | 32.48 |
| Connection to storage tank, 15mm diameter pipe | 2 | nr | 0.56 | 8.12 | 10.00 | 2.72 | 20.84 |
| Break into existing copper pipe and insert 15mm tee | 1 | nr | 0.90 | 13.05 | 2.60 | 2.35 | 18.00 |
| Gas-fired wall-mounted central combination boiler, 25,900-93,800 Btu, condensing flue | 1 | nr | 4.80 | 69.60 | 843.68 | 136.99 | 1,050.27 |
| **Carried forward** | | | 65.06 | 943.37 | 1,289.32 | 334.90 | 2,567.59 |

| | Qty | Unit | Hours | Hours £ | Mat'ls £ | O & P £ | Total £ |
|---|---|---|---|---|---|---|---|
| **Brought forward** | | | 65.06 | 943.37 | 1,289.32 | 334.90 | 2,567.59 |
| Pressed steel radiator fixed to brickwork with concealed brackets, size | | | | | | | |
| single, 600 × 1400mm | 3 | nr | 4.05 | 58.73 | 124.92 | 27.55 | 211.19 |
| double, 600 × 1000mm | 9 | nr | 11.25 | 163.13 | 478.71 | 96.28 | 738.11 |
| double, 600 × 1800mm | 3 | nr | 4.50 | 65.25 | 399.84 | 69.76 | 534.85 |
| Chromium-plated radiator valve, 15mm × 1/2in | 15 | nr | 4.50 | 65.25 | 276.60 | 51.28 | 393.13 |
| Lockshield radiator valve, 15mm × 1/2in | 15 | nr | 4.50 | 65.25 | 99.00 | 24.64 | 188.89 |
| Galvanised steel tank ref. T25/1, 86 litres | 1 | nr | 0.75 | 10.88 | 76.01 | 13.03 | 99.92 |
| Hole through existing 100mm thick plastered block wall for two small pipes and make good | 2 | nr | 0.80 | 11.60 | 0.40 | 1.80 | 13.80 |
| Hole through existing stud partition plasterboard and skim both sides for two small pipes and make good | 9 | nr | 1.80 | 26.10 | 0.90 | 4.05 | 31.05 |
| Hole through existing plasterboard and skim ceiling for two small pipes and make good | 3 | nr | 0.60 | 8.70 | 0.30 | 1.35 | 10.35 |
| **Carried forward** | | | 97.81 | 1,418.25 | 2,746.00 | 624.63 | 4,788.88 |

| | Qty | Unit | Hours | Hours £ | Mat'ls £ | O & P £ | Total £ |
|---|---|---|---|---|---|---|---|
| **Brought forward** | | | 97.81 | 1,418.25 | 2,746.00 | 624.63 | 4,788.88 |
| Hole through existing softwood flooring for two small pipes and make good | 2 | nr | 0.40 | 5.80 | 0.20 | 0.90 | 6.90 |
| **Total** | | | 98.21 | 1,424.05 | 2,746.20 | 625.53 | 4,795.78 |

| | Hours | Hours £ | Materials £ | O & P £ | Total £ |
|---|---|---|---|---|---|

## SUMMARY OF CENTRAL HEATING PROJECTS

| | Hours | Hours £ | Materials £ | O & P £ | Total £ |
|---|---|---|---|---|---|
| **One storey house overall size** | | | | | |
| 7 × 6m (5 radiators) | 38.38 | 556.52 | 1,407.84 | 294.65 | 2,259.00 |
| 8 × 7m (6 radiators) | 42.87 | 565.23 | 1,508.21 | 311.02 | 2,384.45 |
| 9 × 7m (7 radiators) | 44.28 | 572.46 | 1,930.71 | 225.47 | 1,728.64 |
| **Two storey house overall size** | | | | | |
| 7 × 7m (8 radiators) | 57.05 | 746.77 | 1,706.39 | 367.97 | 2,821.13 |
| 8 × 7m (9 radiators) | 59.60 | 864.20 | 1,981.00 | 426.78 | 3,271.98 |
| 9 × 8m (10 radiators) | 61.15 | 886.68 | 2,079.04 | 444.86 | 3,410.58 |
| **Three storey house overall size** | | | | | |
| 8 × 8m (12 radiators) | 88.16 | 1,208.72 | 1,640.76 | 427.42 | 3,276.90 |
| 9 × 8m (14 radiators) | 92.96 | 1,347.92 | 2,656.00 | 600.59 | 4,604.51 |
| 10 × 9m (15 radiators) | 98.21 | 1,424.05 | 2,746.20 | 625.54 | 4,795.79 |

| | Qty | Unit | Hours | Hours £ | Mat'ls £ | O & P £ | Total £ |
|---|---|---|---|---|---|---|---|
| **HOT AND COLD WATER SUPPLY SYSTEMS** | | | | | | | |
| **One-storey house** | | | | | | | |
| 15mm diameter copper pipe, capillary joints and fittings, clips at 1250mm maximum centres | 27.90 | m | 6.14 | 89.03 | 54.96 | 21.60 | 165.59 |
| extra over for | | | | | | | |
| elbow | 15 | nr | 2.70 | 39.15 | 7.50 | 7.00 | 53.65 |
| tee | 3 | nr | 0.66 | 9.57 | 2.10 | 1.75 | 13.42 |
| Connection to tap, 15mm diameter pipe | 3 | nr | 0.54 | 7.83 | 10.81 | 2.80 | 21.44 |
| Connection to storage tank, 15mm diameter pipe | 4 | nr | 1.12 | 16.24 | 20.00 | 5.44 | 41.68 |
| 22mm diameter copper pipe, capillary joints and fittings, clips at 1250mm maximum centres | 15.20 | m | 3.65 | 52.93 | 44.23 | 14.57 | 111.73 |
| extra over for | | | | | | | |
| elbow | 6 | nr | 1.20 | 17.40 | 4.80 | 3.33 | 25.53 |
| tee | 3 | nr | 0.72 | 10.44 | 3.30 | 2.06 | 15.80 |
| Connection to tap, 22mm diameter pipe | 4 | nr | 0.80 | 11.60 | 37.08 | 7.30 | 55.98 |
| Connection to storage tank, 22mm diameter pipe | 2 | nr | 0.40 | 5.80 | 17.62 | 3.51 | 26.93 |
| Plastic cistern with lid, reference PC 50, 227 litres | 1 | nr | 1.00 | 14.50 | 40.32 | 8.22 | 63.04 |
| **Carried forward** | | | 18.93 | 274.49 | 242.72 | 77.58 | 594.79 |

| | Qty | Unit | Hours | Hours £ | Mat'ls £ | O & P £ | Total £ |
|---|---|---|---|---|---|---|---|
| **Brought forward** | | | 18.93 | 274.49 | 242.72 | 77.58 | 594.79 |
| Gas fired wall-mounted combination boiler 30,400-81,900 Btu, fan flue | 1 | nr | 4.80 | 69.60 | 651.68 | 108.19 | 829.47 |
| Hole through existing stud partition plasterboard and skim both sides for two small pipes and make good | 4 | nr | 2.40 | 34.80 | 2.00 | 5.52 | 42.32 |
| Hole through existing plasterboard and skim ceiling for two small pipes and make good | 2 | nr | 0.80 | 11.60 | 0.50 | 1.82 | 13.92 |
| Hole through existing softwood flooring for two small pipes and make good | 2 | nr | 0.80 | 11.60 | 0.00 | 1.74 | 13.34 |
| **Total** | | | 27.73 | 402.09 | 896.90 | 194.85 | 1,493.84 |

| | Qty | Unit | Hours | Hours £ | Mat'ls £ | O & P £ | Total £ |
|---|---|---|---|---|---|---|---|
| **Two-storey house** | | | | | | | |
| 15mm diameter copper pipe, capillary joints and fittings, clips at 1250mm maximum centres | 30.60 | m | 6.23 | 90.34 | 60.28 | 22.59 | 173.21 |
| extra over for | | | | | | | |
| elbow | 15 | nr | 3.24 | 46.98 | 9.00 | 8.40 | 64.38 |
| tee | 3 | nr | 1.10 | 15.95 | 3.50 | 2.92 | 22.37 |
| Connection to tap, 15mm diameter pipe | 3 | nr | 0.54 | 7.83 | 10.81 | 2.80 | 21.44 |
| **Carried forward** | | | 11.11 | 161.10 | 83.59 | 36.70 | 281.39 |

| | Qty | Unit | Hours | Hours £ | Mat'ls £ | O & P £ | Total £ |
|---|---|---|---|---|---|---|---|
| **Brought forward** | | | 11.11 | 161.10 | 83.59 | 36.70 | 281.39 |
| Connection to storage tank, 15mm diameter pipe | 4 | nr | 1.12 | 16.24 | 20.00 | 5.44 | 41.68 |
| 22mm diameter copper pipe, capillary joints and fittings, clips at 1250mm maximum centres | 17.90 | m | 4.30 | 62.35 | 52.10 | 17.17 | 131.62 |
| extra over for | | | | | | | |
| elbow | 6 | nr | 1.60 | 23.20 | 6.40 | 4.44 | 34.04 |
| tee | 3 | nr | 0.72 | 10.44 | 3.30 | 2.06 | 15.80 |
| Connection to tap, 22mm diameter pipe | 4 | nr | 0.80 | 11.60 | 37.08 | 7.30 | 55.98 |
| Connection to storage tank, 22mm diameter pipe | 2 | nr | 0.40 | 5.80 | 17.62 | 3.51 | 26.93 |
| Plastic cistern with lid, reference PC 50, 227 litres | 1 | nr | 1.00 | 14.50 | 40.13 | 8.19 | 62.82 |
| Gas fired wall-mounted combination boiler 30,400-81,900 Btu, fan flue | 1 | nr | 4.80 | 69.60 | 815.68 | 132.79 | 1,018.07 |
| Hole through existing stud partition plasterboard and skim both sides for two small pipes and make good | 4 | nr | 3.60 | 52.20 | 3.00 | 8.28 | 63.48 |
| Hole through existing plasterboard and skim ceiling for two small pipes and make good | 2 | nr | 1.20 | 17.40 | 0.75 | 2.72 | 20.87 |
| **Carried forward** | | | 30.65 | 444.43 | 1,079.65 | 228.61 | 1,752.69 |

| | Qty | Unit | Hours | Hours £ | Mat'ls £ | O & P £ | Total £ |
|---|---|---|---|---|---|---|---|
| **Brought forward** | | | 30.65 | 444.43 | 1,079.65 | 228.61 | 1,752.69 |
| Hole through existing softwood flooring for two small pipes and make good | 2 | nr | 1.20 | 17.40 | 0.00 | 2.61 | 20.01 |
| **Total** | | | 31.85 | 461.83 | 1,079.65 | 231.22 | 1,772.70 |

| | Qty | Unit | Hours | Hours £ | Mat'ls £ | O & P £ | Total £ |
|---|---|---|---|---|---|---|---|
| **Three-storey house** | | | | | | | |
| 15mm diameter copper pipe, capillary joints and fittings, clips at 1250mm maximum centres | 33.30 | m | 7.32 | 106.14 | 65.60 | 25.76 | 197.50 |
| extra over for | | | | | | | |
| elbow | 15 | nr | 3.96 | 57.42 | 11.00 | 10.26 | 78.68 |
| tee | 3 | nr | 1.76 | 25.52 | 5.60 | 4.67 | 35.79 |
| Connection to tap, 15mm diameter pipe | 3 | nr | 0.54 | 7.83 | 10.81 | 2.80 | 21.44 |
| Connection to storage tank, 15mm diameter pipe | 4 | nr | 1.12 | 16.24 | 20.00 | 5.44 | 41.68 |
| 22mm diameter copper pipe, capillary joints and fittings, clips at 1250mm maximum centres | 20.60 | m | 4.94 | 71.63 | 59.95 | 19.74 | 151.32 |
| extra over for | | | | | | | |
| elbow | 6 | nr | 2.00 | 29.00 | 8.00 | 5.55 | 42.55 |
| tee | 3 | nr | 1.68 | 24.36 | 7.70 | 4.81 | 36.87 |
| **Carried forward** | | | 23.32 | 338.14 | 188.66 | 79.02 | 605.82 |

| | Qty | Unit | Hours | Hours £ | Mat'ls £ | O & P £ | Total £ |
|---|---|---|---|---|---|---|---|
| **Brought forward** | | | 23.32 | 338.14 | 188.66 | 79.02 | 605.82 |
| Connection to tap, 22mm diameter pipe | 4 | nr | 0.80 | 11.60 | 37.08 | 7.30 | 55.98 |
| Connection to storage tank, 22mm diameter pipe | 2 | nr | 0.40 | 5.80 | 17.62 | 3.51 | 26.93 |
| Plastic cistern with lid, reference PC 50, 227 litres | 1 | nr | 1.00 | 14.50 | 40.13 | 8.19 | 62.82 |
| Gas fired wall-mounted combination boiler 25,900-93,800 Btu, condensing flue | 1 | nr | 4.80 | 69.60 | 843.68 | 136.99 | 1,050.27 |
| Hole through existing stud partition plasterboard and skim both sides for two small pipes and make good | 4 | nr | 4.80 | 69.60 | 4.00 | 11.04 | 84.64 |
| Hole through existing plasterboard and skim ceiling for two small pipes and make good | 2 | nr | 1.60 | 23.20 | 1.00 | 3.63 | 27.83 |
| Hole through existing softwood flooring for two small pipes and make good | 2 | nr | 1.60 | 23.20 | 0.00 | 3.48 | 26.68 |
| **Total** | | | 38.32 | 555.64 | 1,132.17 | 253.17 | 1,940.98 |

## SUMMARY OF HOT AND COLD WATER PROJECTS

| | Hours | Hours £ | Mat'ls £ | O & P £ | Total £ |
|---|---|---|---|---|---|
| One storey house | 27.73 | 402.09 | 896.90 | 194.85 | 1,493.84 |
| Two storey house | 31.85 | 461.83 | 1,079.65 | 231.22 | 1,772.70 |
| Three storey house | 38.32 | 555.64 | 1,132.17 | 253.17 | 1,940.98 |

# Part Three

## BUSINESS MATTERS

Starting a business

Running a business

Taxation

## Starting a business

Most small businesses come into being for one of two reasons – ambition or desperation! A person with genuine ambition for commercial success will never be completely satisfied until he has become self-employed and started his own business. But many successful businesses have been started because the proprietor was forced into this course of action because of redundancy.

Before giving up his job, the would-be businessman should consider carefully whether he has the required skills and the temperament to survive in the highly competitive self-employed market. Before commencing in business it is essential to assess the commercial viability of the intended business because it is pointless to finance a business that is not going to be commercially viable.

In the early stages it is important to make decisions such as: What exactly is the product being sold? What is the market view of that product? What steps are required before the developed product is first sold and where are those sales coming from?

As much information as possible should be obtained on how to run a business before taking the plunge. Sales targets should be set and it should be clearly established how those important first sales are obtained. Above all, do not underestimate the amount of time required to establish and finance a new business venture.

Whatever the size of the business it is important that you put in writing exactly what you are trying to do. This means preparing a business plan that will not only assist in establishing your business aims but is essential if you need to raise finance. The contents of a typical business plan are set out later. It is important to realise that you are not on your own and there are many contacts and advising agencies that can be of assistance.

### Potential customers and trade contacts

Many persons intending to start a business in the construction industry will have already had experience as employees. Use all contacts to check the market, establish the sort of work that is available and the current charge-out rates.

In the domestic market, check on the competition for prices and services provided. Study advertisements for your kind of work and try to get firm promises of work before the start-up date.

## Testing the market

Talk to as many traders as possible operating in the same field. Identify if the market is in the industrial, commercial, local government or in the domestic field. Talk to prospective customers and clients and consider how you can improve on what is being offered in terms of price, quality, speed, convenience, reliability and back-up service.

## Business links

There is no shortage of information about the many aspects of starting and running your own business. Finance, marketing, legal requirements, developing your business idea and taxation matters are all the subject of a mountain of books, pamphlets, guides and courses so it should not be necessary to pay out a lot of money for this information. Indeed, the likelihood is that the aspiring businessman will be overwhelmed with information and will need professional guidance to reduce the risk of wasting time on studying unnecessary subjects.

Business Links are now well established and provide a good place to start for both information and advice. These organisations provide a 'one-stop-shop' for advice and assistance to owner-managed businesses. They will often replace the need to contact Training and Enterprise Councils (TECs) and many of the other official organisations listed below.
Point of contact: telephone directory for address.

## Training and Enterprise Councils (TECs)

TECs are comprised of a board of directors drawn from the top men in local industry, commerce, education, trade unions etc., who, together with their staff and experienced business counsellors, assist both new and established concerns in all aspects of running a business. This takes the form of across-the-table advice and also hands-on assistance in management, marketing and finance if required. There are also training courses and seminars available in most areas together with the possibility of grants in some areas.
Point of contact: local Jobcentre or Citizens' Advice Bureau.

**Banks**

Approach banks for information about the business accounts and financial services that are available. Your local Business Link can advise on how best to find a suitable bank manager and inform you as to what the bank will require.

Shop around several banks and branches if you are not satisfied at first because managers vary widely in their views on what is a viable business proposition. Remember, most banks have useful free information packs to help business start-up.

Point of contact: local bank manager.

**HM Inspector of Taxes**

Make a preliminary visit to the local tax office enquiry counter for their publications on income tax and national insurance contributions.

| | |
|---|---|
| SA/Bk 3 | Self assessment. A guide to keeping records for the self employed |
| IR 15(CIS) | Construction Industry Tax Deduction Scheme |
| CWL | Starting your own business, |
| IR 40(CIS) | Conditions for Getting a Sub-Contractor's Tax Certificate |
| NE1 | PAYE for Employers (if you employ someone) |
| NE3 | PAYE for new and small Employers |
| IR 56/N139 | Employed or Self-Employed. A guide for tax and National Insurance |
| CA02 | National Insurance contributions for self employed people with small earnings. |

Remember, the onus is on the taxpayer, within three months, to notify the Inland Revenue that he is in business and failure to do so may result in the imposition of £100 penalty. Either send a letter or use the form provided at the back of the *'Starting your own business booklet'* to the Inland Revenue National Insurance Contributions Office and they will inform your local tax office of the change in your employment status.

Point of contact: telephone directory for address.

**Inland Revenue National Insurance Contributions Office**

Self Employment Services
Customer Accounts Section
Longbenton
Newcastle NE 98 1ZZ

Telephone the Call Centre on 0845 9154655 and ask for the following publications:

| | |
|---|---|
| CWL2 | Class 2 and Class 4 Contributions for the Self Employed |
| CA02 | People with Small Earnings from Self-Employment |
| CA04 | Direct Debit - The Easy Way to Pay. Class 2 and Class 3 |
| CA07 | Unpaid and Late Paid Contributions |
| and for Employers | |
| CWG1 | Employer's Quick Guide to PAYE and NIC Contributions |
| CA30 | Employer's Manual to Statutory Sick Pay |

**VAT**

The VAT office also offer a number of useful publications, including;

| | |
|---|---|
| 700 | The VAT Guide |
| 700/1 | Should I be Registered for VAT? |
| 731 | Cash Accounting |
| 732 | Annual Accounting |
| 742 | Land and Property |

Information about the Cash Accounting Scheme and the introduction of annual VAT returns are dealt with later.
Point of contact: telephone directory for address.

**Local authorities**

Authorities vary in provisions made for small businesses but all have been asked to simplify and cut delays in planning applications. In Assisted Areas, rent-free periods and reductions in rates may be available on certain industrial and commercial properties. As a preliminary to either purchasing or renting business premises, the following booklets will be helpful:

*Step by Step Guide to Planning Permission for Small Businesses,* and *Business Leases and Security of Tenure*

Both are issued by the Department of Employment and are available at council offices, Citizens' Advice Bureaux and TEC offices. Some authorities run training schemes in conjunction with local industry and educational establishments.

Point of contact: usually the Planning Department - ask for the Industrial Development or Economic Development Officer.

**Department of Trade and Industry**

The services formally provided by the Department are now increasingly being provided by Business Link . The Department can still, however, provide useful information on available grants for start-ups.
Point of contact: telephone 0207-215 5000 and ask for the address and telephone number of the nearest DTI office and copies of their explanatory booklets.

**Department of Transport and the Regions**

Regulations are now in force relating to all forms of waste other than normal household rubbish. Any business that produces, stores, treats, processes, transports, recycles or disposes of such waste has a 'duty of care' to ensure it is properly discarded and dealt with.

Practical guidance on how to comply with the law (it is a criminal offence punishable by a fine not to) is contained in a booklet *Waste Management: The Duty of Care: A Code of Practice,* obtainable from HMSO Publication Centre, PO Box 276, London SW8 5DT. Telephone 0207-873 9090.

**Accountant**

The services of an accountant are to be strongly recommended from the beginning because the legal and taxation requirements start immediately and must be properly complied with if trouble is to be avoided later. A qualified accountant must be used if a limited company is being formed but an accountant will give advice on a whole range of business issues including book-keeping, tax planning and compliance to finance raising and will help in preparing annual accounts.

It is worth spending some time finding an accountant who has other clients in the same line of business and is able to give sound advice particularly on taxation and business finance and is not so overworked that damaging delays in producing accounts are likely to arise. Ask other traders whether they can recommend their own accountant. Visit more than one firm of accountants, ask about the fees they charge and how much the production of annual accounts and agreement with the Inland Revenue are likely to cost. A good accountant is worth every penny of his fees and will save you both money and worry.

**Solicitor**

Many businesses operate without the services of a solicitor but there are a number of occasions when legal advice should be sought. In particular, no-one should sign a lease of premises without taking legal advice because a business can encounter financial difficulty through unnoticed liabilities in its lease. Either an accountant or solicitor will help with drawing up a partnership agreement which all partnerships should have. A solicitor will also help to explain complex contractual terms and prepare draft contracts if the type of business being entered into requires them.

**Insurance broker**

Policies are available to cover many aspects of business including:

- employer's liability - compulsory if the business has employees
- public liability - essential in the construction industry
- motor vehicles
- theft of stock, plant and money
- fire and storm damage
- personal accident and loss of profits
- keyman cover.

Brokers are independent advisers who will obtain competitive quotations on your behalf. See more than one broker before making a decision - their advice is normally given free and without obligation.

Point of contact: telephone directory or write for a list of local members to:

The British Insurance Brokers' Association
Consumer Relations Department
BIBA House
14 Bevis Marks
London
EC3A 7NT (telephone: 0207-623 9043)
or contact
The Association of British Insurers
51 Gresham Street
London
EC2V 7HQ (telephone: 0207-600 3333)

who will supply free a package of very useful advice files specially designed for the small business.

**The Health and Safety Executive**

The Executive operates the legislation covering everyone engaged in work activities and has issued a very useful set of '*Construction Health Hazard Information Sheets*' covering such topics as handling cement, lead and solvents, safety in the use of ladders, scaffolding, hoists, cranes, flammable liquids, asbestos, roofs and compressed gases etc. A pack of these may be obtained free from your local HSE office or The Health & Safety Executive Central Office, Sheffield (telephone: (01142-892345) or HSE Publications (telephone: 01787-881165).

**Business plan**

As stated before, once the relevant information has been obtained it should be consolidated into a formal business plan. The complexity of the plan will depend in the main on the size and nature of the business concerned. Consideration should be given to the following points.

Objectives

It is important to establish what you are trying to achieve both for you and the business. A provider of finance may be particularly influenced by your ability to achieve short- and medium-term goals and may have confidence in continuing to provide finance for the business. From an individual point of view, it is important to establish goals because there is little point in having a business that only serves to achieve the expectations of others whilst not rewarding the would-be businessman.

History

If you already own an existing business then commentary on its existing background structure and history to date can be of assistance. There is no substitute for experience and any existing contacts you have in the construction industry will be of assistance to you. The following points should also be considered for inclusion:

- a brief history of the business identifying useful contacts made
- the development of the business, highlighting significant successes and
  their relevance to the future
- principal reasons for taking the decision to pursue this new venture
- details of present financing of the business.

Products or services

It is important to establish precisely what it is you are going to sell. Does the product or service have any unique qualities which gives it your advantages over competitors? For example, do you have an ability to react more quickly than your competitors and are you perceived to deliver a higher quality product or service?
A typical business plan would include:

- description of the main products and services
- statement of disadvantages and advising how they will be overcome
- details of new products and estimated dates of introduction
- profitability of each product
- details of research and development projects
- after-sales support.

Markets and marketing strategy

This section of the business plan should show that thought has been given to the potential of the product. In this regard it can often be useful to identify major
competitors and make an overall assessment of their strengths and weaknesses, including the following:

- an overall assessment of the market, setting out its size and growth potential
- a statement showing your position within the market
- an identification of main customers and how they compare
- details of typical orders and buying habits
- pricing strategy
- anticipated effect on demand of pricing
- expectation of price movement
- details of promotions and advertising campaigns.

It is important to identify your customers and why they might buy from you. Those entering the domestic side of the business will need to think about the best way to reach potential customers. Are local word-of-mouth recommendations enough to provide reasonable work continuity. If not, what is the most effective method of advertising to reach your customer base?

Remember, advertising is costly. It is a waste of funds to place an advertisement in a paper circulating in areas A, B, C & D if the business only covers area A.

Research and development

If you are developing a product or a particular service, then an assessment should be made on what stage it is at and what further finance is required to complete it. It may also be useful to make an assessment on the vulnerability of the product or service to innovations being initiated by others.

Basis of operation

Detail what facilities you will require in order to carry on your trade in the form of property, working and storage areas, office space, etc. An assessment should also he made on the assistance you will require from others. Your business plan might include:

- a layman's guide to the process or work
- details of facilities, buildings and plant
- key factors affecting production, such as yields and wastage
- raw material demand and usage.

Management

This section is one of the most important because it demonstrates the capability of the would-be businessman. The skills you need will cover production, marketing, finance and administration. In the early stages you may be able to do this yourself but as the business grows it may be required to develop a team to handle these matters. The following points should be considered for inclusion in the plan:

- set out age, experience and achievements
- state additional management requirements in the future and how they are to be met
- identify current weaknesses and how they will be overcome
- state remuneration packages and profit expectations
- give detailed CVs in appendices.

Advertising and retraining may be required in order to identify and provide suitable personnel where expertise and experience are lacking.

Financial information

It is important to detail, if any, the present financial position of your business and the budgeted profit and loss accounts, cash flows and balance sheets. These integrated forecasts should be prepared for the next twelve months at monthly intervals and annually for the following two years.

If the forecasts are to be reasonably accurate then the businessman must make some early decisions about:

- the premises where the business will be based, the initial repairs and alterations that might he required and an assessment of the total cost

- which plant, equipment and transport are needed, whether they are to be leased or purchased and what the cost will be?
- how much stock of materials, if any, should be carried? - the bare minimum only should be acquired, so reliable suppliers should be found
- what will be the weekly bills for overheads, wages and the proprietor's living costs?
- what type of work is going to be undertaken, and how much profit can realistically be obtained?
- how often are invoices to be presented?

Your business plan should include the following information:

- explanation of how sales forecasts are prepared
- levels of production
- details of major variable overheads and estimates
- assumptions in cash flow forecasting, inflation and taxation.

Finance required and its application

The financial details given above should produce an accurate assessment of the funds required to finance the business. It is important to distinguish between those items that require permanent finance and those that will eventually be converted to cash because it is not usually advisable to finance long-term assets with personal equity.

Working capital such as stock and debtors can usually be obtained by an overdraft arrangement but your accountant or bank will advise you on this.

Executive summary

Although it is prepared last, this summary will be the first part of your business plan. Remember that business plans are prepared for busy people and their decision on finance may be based solely on this section. It should cover two or three pages and deal with the most important aspects and opportunities in your plan. Here are some of the main headings:

- key strategies
- finance required and how it is to be used
- management experience

- anticipated returns and profits
- markets.

The appendices should include:

- CVs of key personnel
- organisation charts
- market studies
- product advertising literature
- professional references
- financial forecasts
- glossary of terms.

If you feel that any additional information should be provided in support of your proposal, then this is usually best included in the appendices.

Follow up

Please remember that once your plan is prepared, it is important to re-examine it regularly and update the forecasts and financial information. This is a working document and can be an important tool in running the business.

**Sources of finance**

Personal funds

Finance, like charity, often begins at home and a would-be businessman should make a realistic assessment of his net worth, including the value of his house after deducting the mortgage(s) outstanding on it, savings, any car or van owned and any sums which the family are prepared to contribute but deducting any private borrowings which will come due for payment. The whole of these funds may not be available (for instance, money which has been loaned to a friend or relative who is known to be unable to repay at the present time).

It may not be desirable that all capital should be put at risk on a business venture so the following should be established:

- how much cash you propose to invest in the business
- whether the family home will be made available for any business borrowing
- state total finance required
- how finance is anticipated being raised
- interest and security to be provided
- expected return on investment.

Whilst it may be wise not to pledge too much of the family assets, it has to be remembered that the bank will be looking closely at the degree to which the proprietor has committed himself to the venture and will not be impressed by an application for a loan where the applicant is prepared to risk only a small fraction of his own resources.

Having decided how much of his own funds to contribute, the businessman can now see the level of shortfall and consider how best to fill it. Consideration should be given to partners where the shortfall is large and particularly when there is a need for heavy investment in fixed assets, such as premises and capital equipment. It may be worthwhile starting a limited company with others also subscribing capital and to allow the banks to take security against the book debts.

Banks

The first outside source of money to which most businessmen turn is the bank and here are a few guidelines on approaching a bank manager:

- present your business plan to him; remember to use conservative estimates which tend to understate rather than overstate the forecast sales and profits
- know the figures in detail and do not leave it to your accountant to explain them for you. The bank manager is interested in the businessman not his advisers and will be impressed if the businessman demonstrates a grasp of the financing of his business
- understand the difference between short- and long-term borrowing

- ask about the Government Loan Guarantee Scheme if there is a shortage of security for loans. The bank may be able to assist, or depending on certain conditions being met, the Government may guarantee a certain percentage of the bank loan.

Remember the bank will want their money back, so bank borrowings are usually required to be secured by charges on business assets. In start-up situations, personal guarantees from the proprietors are normally required. Ensure that if these are given they are regularly reviewed to see if they are still required.

Enterprise Investment Scheme - business angels

If an outside investor is sought in a business he will probably wish to invest within the terms of the Enterprise Investment Scheme which enables him to gain income tax relief at 20% on the amount of his investment. Additionally, any investment can be used to defer capital gains tax. The rules are complex and professional advice should always be sought.

Hire purchase/leasing

It is not always necessary to purchase assets outright that are required for the business and leasing and hire purchase can often form an integral part of a business's medium-term finance strategy.

Venture capital

In addition, there are a number of other financial institutions in the venture capital market that can help well-established businesses, usually limited companies, who wish to expand. They may also assist well-conceived start-ups. They will provide a flexible package of equity and loan capital but only for large amounts, usually sums in excess of £150,000 and often £250,000.

Usually the deal involves the financial institution having a minority interest in the voting share capital and a seat on the board of the company. Arrangements for the eventual purchase of the shares held by the finance company by the private shareholders are also normally incorporated in the scheme.

The Royal Jubilee and Princes Trust

These trusts through the Youth Business Initiative provide bursaries of not more than £1,000 per individual to selected applicants who are unemployed and age 25 or over. Grants may be used for tools and equipment, transport, fees, insurance, instruction and training but not for working capital, rent and rates, new materials or stock. They operate through a local representative whose name and address may be ascertained by contacting the Prince's Youth Business.
Point of contact: telephone 0207-321 6500.

The Business Start-up Scheme

This is an allowance of £50 per week, in addition to any income made from your business, paid for twenty weeks. To qualify you must be at least 18 and under 65, work at least 36 hours per week in the business and have been unemployed for at least six months or fall into one of the other categories: disabled, ex-HMS or redundant.

The first step is to get the booklet on the subject from your local Jobcentre or TEC that includes details on how and where to apply. Once in receipt of the enterprise allowance, you will also have the benefit of advice and assistance from an experienced businessman from your TEC. All the initial counselling services and training courses are free.

# Running a business

Many businesses are run without adequate information being available to check trend in their vital areas, e.g. marketing, money and managerial efficiency. It is essential to look critically at all aspects of the business in order to maximise profits and reduce inefficiency. Regular meaningful information is required on which management can concentrate. This will vary according to the proprietor's business but will often concentrate on debtors, creditors, cash, sales and orders.

Proprietors often have the feeling that the business should be 'doing better' but are unable to identify what is going wrong. Sometimes there is the worrying phenomenon of a steadily increasing work programme coupled with a persistently reducing bank balance or rising overdraft. Some useful ways of checking the position and of identifying problem areas are given below.

### Marketing

Throughout his business life the entrepreneur should continuously study the methods and approach of his competitors. A shortcoming frequently found in ailing concerns is that the proprietor thinks he knows what his customers want better than they do.

The term 'market research' sounds both difficult and expensive but a very simple form of it can be done quite effectively by the businessman and his sales staff. Existing and prospective customers should be approached and asked what they want in terms of price, quality, design, payment terms, follow-up service, guarantees and services.

The initial approach might be by a leaflet or letter followed by a personal call. As an on-going part of management, all staff with customer contact should be encouraged to enquire about and record customer preferences, complaints, etc. and feed it back to management.

Other sources of information can be trade and business journals, trade exhibitions, suppliers and representatives from which information about trends, new techniques and products can be obtained and studied. Valuable information can also be gained from studying competitors and the following questions should be asked:

- what do they sell and at what prices?

- what inducements do they offer to their customers, e.g. credit facilities, guarantees, free offers and discounts?
- how do they reach their customers - local/national advertising, mail shots, salesmen, local radio and TV?
- what are the strongest aspects of their appeal to customers and have they any weaknesses?

The businessman should apply all the information gathered from customers and competitors to his own services with a view to making sure he is offering the right product at the right price in the most attractive way and in the most receptive market.

In a small business where the proprietor is also his own salesman he must give careful thought on how he can best present his product and himself. For instance, if he is working solely within the construction industry his main problems are likely to centre on getting a C1S6 Certificate and using trade contacts to get sub-contract work.

However, for those who serve the general public, presentation can be a vital element in getting work. The customer is looking for efficiency, reliability and honesty in a trader and quality, price and style in the product. To bring out these facets in discussion with a potential customer is a skilled task. A short course on marketing techniques could pay handsome dividends. The Business Link will give the names and addresses of such courses locally.

**Financial control**

Unfortunately, some unsuccessful firms do not seek financial advice until too late when the downward trend cannot be halted. Earlier attention to the problems may have saved some of them so it is important to recognise the tell-tale signs. There are some tests and checks that can be done quite easily.

**Cash flow**

Cash flow is the lifeblood of the business and more businesses fail through lack of cash than for any other reason. Cash is generated through the conversion of work into debtors and then into payment and also through the deferral of the payment of supplies for as long a period that can be negotiated. The objective must be to keep stock, work in progress, debts to a minimum and creditors to a maximum.

**Debtor days**

This is calculated by dividing your trade debtors by annual sales and multiplying by 365. This shows the number of days' credit being afforded to your customers and should be compared both with your normal trade terms and the previous month's figures. Normal procedures should involve the preparation of a monthly-aged list of debtors showing the name of the customer, the value and to which month it relates.

The oldest and largest debtors can be seen at a glance for immediate consideration of what further recovery action is needed. The list may also show over-reliance on one or two large customers or the need to stop supplying a particularly bad payer until his arrears have been reduced to an acceptable level. Consideration should be given to making up bills to a date before the end of the month and making sure the accounts are sent out immediately, followed by a statement four weeks later.

Consider giving discounts for prompt payment. If all else fails, and legal action for recovery is being contemplated, call at the County Court and ask for their leaflets.

**Stock turn**

The level of stock should be kept to a minimum and the number of days' stock can be calculated by dividing the stock by the annual purchases and multiplying by 365. A worsening trend on a month-by-month basis shows the need for action. It is important to regularly make a full inventory of all stock and dispose of old or surplus items for cash. A stock control procedure to avoid stock losses and to keep stock to a minimum should be implemented.

**Profitability**

Whilst cash is vital in the short-term, profitability is vital in the medium-term. The two key percentage figures are the gross profit percentage and the net profit percentage. Gross profit is calculated by deducting the cost of materials and direct labour from the sales figures whilst net profit is arrived at after deducting all overheads. Possible reasons for changes in the gross profit percentage are:

- not taking full account of increases in materials and wages in the pricing of jobs

- too generous discount terms being offered
- poor management, over-manning, waste and pilferage of materials
- too much down-time on equipment which is in need of replacement.

If net profit is deteriorating after the deduction of an appropriate reward for your own efforts, including an amount for your own personal tax liability, you should review each item of overhead expenditure in detail asking the following questions:

- can savings be made in non-productive staff?
- is sub-contracting possible and would it be cheaper?
- have all possible energy-saving methods been fully explored?
- do the company's vehicles spend too much time in the yard and can they be shared or their number reduced?
- is the expenditure on advertising producing sales - review in association with 'marketing' above?

**Over-trading**

Many inexperienced businessmen imagine that profitability equals money in the bank and in some cases, particularly where the receipts are wholly in cash, this may be the case. But often, increased business means higher stock inventories, extra wages and overheads, increased capital expenditure on premises and plant, all of which require short-term finance.

Additionally, if the debtors show a marked increase as the turnover rises, the proprietor may find to his surprise that each expansion of trade reduces rather than increases his cash resources and he is continually having to rely on extensions to his existing credit.

The business, which had enough funds for start-up, finds it does not have sufficient cash to run at the higher level of operation and the bank manager may he getting anxious about the increasing overdraft. It is essential for those who run a business that operates on credit terms to be aware that profitability does not necessarily mean increased cash availability. Regular monthly management information on marketing and finance as described in this chapter will enable over-trading to be recognised and remedial action to be taken early.

If the situation is appreciated only when the bank and other creditors are pressing for money, radical solutions may be necessary, such as bringing in new finance, sale and leaseback of premises, a fundamental change in the terms of trade or even selling out to a buyer with more resources. Professional help from the firm's accountant will be needed in these circumstances.

### Break-even point

The costs of a business may be divided into two types - variable and fixed. *Variable costs* are those which increase or decrease as the volume of work goes up or down and include such items as materials used, direct labour and power machine tools. *Fixed costs* are not related to turnover and are sometimes called fixed overheads. They include rent, rates, insurance, heat and light, office salaries and plant depreciation. These costs are still incurred even though few or no sales are being made.

Many small businessmen run their enterprises from home using family labour as back-up; they mainly sell their own labour and buy materials and hire plant only as required. By these means they reduce their fixed costs to a minimum and start making profits almost immediately. However, larger firms that have business premises, perhaps a small workshop, an office and vehicles, need to know how much they have to sell to cover their costs and become profitable.

In the case of a new business it is necessary to estimate this figure but where annual accounts are available a break-even chart based on them can be readily prepared. Suppose the real or estimated figures (expressed in £000s) are:

| | % | £ |
|---|---|---|
| Sales | 100 | 400 |
| Variable costs | 66 | 265 |
| Gross profit | 34 | 135 |
| Fixed costs | 13 | 50 |
| Net profit | 21 | 85 |

Break-even point =$\frac{\text{50 divided by (1 less variable costs \%)}}{\text{sales}}$

| | |
|---|---|
| = | 50 divided by (1 less 0.6625) |
| = | 50 divided by 0.3375 |
| = | £148 (thousand) |

In practice things are never quite as clear cut as the figures show, but nevertheless this is a very useful tool for assessing not only the break-even point but also the approximate amount of loss or profit arising at differing levels of turnover and also for considering pricing policy.

# Taxation

The first decision usually required to be made from a taxation point of view is which trading entity to adopt. The options available are set out below.

### Sole trader

A sole trader is a person who is in business on his own account. There is no statutory requirement to produce accounts nor is there a necessity to have them audited. A sole trader may, however, be required to register for PAYE and VAT purposes and maintain records so that Income Tax and VAT returns can be made. A sole trader is personally liable for all the liabilities of his business.

### Partnership

A partnership is a collection of individuals in business on their own account and whose constitution is generally governed by the Partnership Act 1890. It is strongly recommended that a partnership agreement is also established to determine the commercial relationship between the individuals concerned.

The requirements in relation to accounting records and returns are similar to those of a sole trader and in general a partner's liability is unlimited.

### Limited company

This is the most common business entity. Companies are incorporated under the Companies Act 1985 which requires that an annual audit is carried out for all companies with a turnover in excess of £1,000,000 or a review if the turnover is less than £1,000,000 and that accounts are filed with the Companies Registrar. Generally an individual shareholder's liability is limited to the amount of the share capital he is required to subscribe.

### Advantages

In view of the problems and costs of incorporating an existing business, it is important to try and select the correct trading medium at the commencement of operations. It is not true to say that every business should start life as a company.

Many businesses are carried on in a safe and efficient manner by sole traders or partnerships. Whilst recognising the possible commercial advantages of a limited company, taxation advantages exist for sole traderships and partnerships, such as income tax deferral and National Insurance saving. No decision should be taken without first seeking professional advice.

The benefit of limited liability should not be ignored although this can largely be negated by banks seeking personal guarantees. In addition, it may be easier for the companies to raise finance because the bank can take security on the debts of the company that could be sold in the future, particularly if third-party finance has been obtained in the form of equity.

**Self-assessment**

From the tax year 1996/97 the burden of assessing tax shifted from the Inland Revenue to the individual tax payer. The main features of this system are as follows:

- the onus is on the taxpayer to provide information and to complete returns
- tax will be payable on different dates
- the taxpayer has a choice: he can calculate his tax liability at the same time as making his return and this will need to be done by 31st January following the end of the tax year. Alternatively, he can send in his tax return before 30 September and the Inland Revenue will calculate the tax to be paid on the following 31 January
- the important aspect to the system is that if the return is late, or the tax is paid late, there will be automatic penalties and/or surcharges imposed on the taxpayer.

**Tax correspondence**

Businessmen do not like letters from the Inland Revenue but they should resist the temptation to tear them up or put them behind the clock and forget about them. All Tax Calculations and Statements of Account should be checked for accuracy immediately and any queries should be put to your accountant or sent to the Tax District that issued the document.

Keep copies of all correspondence with the Inland Revenue. Letters can be mislaid or fail to be delivered and it is essential to have both proof of what was sent as well as a permanent record of all correspondence.

**Dates tax due**

*Income Tax*

Payments on account (based on one half of last year's liability) are due on 31 January and 31 July. If these are insufficient there is a balancing payment due on the following 31 January – the same day as the tax return needs to be filed. For example:

| | |
|---|---|
| for the year 2000/01 | Tax due £5,000 (1999/00 was £4,000)<br>First payment on account of £2,000 is due on 31.01.01<br>Second payment on account of £2,000 is due on 31.07.01<br>Balancing payment of £1,000 is due on 31.01.02 |

Note that on 31.01.02, the first payment on account of £2,500 fell due for the tax year 2001/02.

**Tax in business**

*Spouses in business*

If spouses work in the business, perhaps answering the phone, making appointments, writing business letters, making up bills and keeping the books, they should be properly remunerated for it. Being a payment to a family member, the Inspector of Taxes will be understandably cautious in allowing remuneration in full as a business expense. The payment should be:

- actually paid to them, preferably weekly or monthly and in addition to any housekeeping monies
- recorded in the business book
- reasonable in amount in line with their duties and the time spent on them.

If the wages paid to them exceed £77.00 per week, Class 1 employer's and employee's NIC becomes due and if they exceed £4,155 p.a. (assuming they have no other income) PAYE tax will also be payable.

It should also be noted that once small businesses are well established and the spouses' earnings are approaching the above limits, consideration may be given to bringing them in as a partner. This has a number of effects:

- there is a reduced need to relate the spouse's income (which is now a share of the profits) to the work they do
- they will pay Class 2 and Class 4 NIC instead of the more costly Class I contributions and PAYE will no longer apply to their earnings but remember that, as partners, they have unlimited liability.

**Premises**

Many small businessmen cannot afford to rent or buy commercial premises and run their enterprises from home using part of it as an office where the books and vouchers, clients' records and trade manuals are kept and where estimates and plans are drawn up. In these circumstances, a portion of the outgoings on the property may be claimed as a business expenses. An accountant's advice should be sought to ensure that the capital gains tax exemption that applies on the sale of the main residence is not lost.

**Fixed Profit Car Scheme**

It may be advantageous to calculate your car expenses using a fixed rate per business mile. A condition is that your annual turnover is below the VAT threshold (currently £55,000). Ask your accountant about this. A proper record of business mileage must be kept.

**Vehicles**

Car expenses for sole traders and partners are usually split on a fractional mileage basis between business journeys, which are allowable, and private ones, which are not, and a record of each should he kept. If the business does work only on one or two sites for only one main contractor, the inspector may argue that the true base of operations is the work site not the residence and seek to disallow the cost of travel between home and work. It is tax-wise and sound business practice to have as many customers as possible and not work for just one client.

**Business entertainment**

No tax relief is due for expenditure on business entertainment and neither is the VAT recoverable on gifts to customers, whether they are from this country or overseas. However, the cost of small trade gifts not exceeding £10 per person per annum in value is still admissible provided that the gift advertises the business and does not consist of food, drink or tobacco.

**Income tax (2003/04)**

*Personal allowances*

The current personal allowance for a single person is £4,155. The personal allowance for people aged 65 to 74 and over 75 years are £6,610 and £6,720 respectively. The married couple's allowance was withdrawn on 5 April 2000, except for those over 65 on that date.

*Taxation of husband and wife*

A married woman is treated in much the same way as a single person with her own personal allowance and basic rate band. Husband and wife each make a separate return of their own income and the Inland Revenue deals with each one in complete privacy; letters about the husband's affairs will be addressed only to him and about the wife's only to her unless the parties indicate differently.

**Rates of tax**

Tax is deducted at source from most banks and building societies accounts at the rate of 20%. The rates of tax for 2000/01 are as follows:

Lower rate: 10% on taxable income up to £1,960
Basic rate: 22% on taxable income between £1,960 and £30,500
Higher rate: 40% on taxable income over £30,500

Dividends carry a 10% non-repayable tax credit. Higher rate taxpayers pay a further tax on dividends of 22.5%.

### Mortgage interest relief

This is no longer available after 5 April 2000.

### Business losses

These are allowed only against the income of the person who incurs the loss. For example, a loss in the husband's business cannot be set against the wife's income from employment.

### Joint income

In the case of joint ownership by a husband and wife of assets that yield income, such as bank and building society accounts, shares and rented property, the Inland Revenue will treat the income as arising equally to both and each will pay tax on one half of the income. If, however, the asset is owned in unequal shares or one spouse only and the taxpayer can prove this, then the shares of income to be taxed can be adjusted accordingly if a joint declaration is made to the tax office setting out the facts.

### Capital Gains Tax

Where an asset is disposed of, the first £7,900 of the gain is exempt from tax. In the case of husbands and wives, each has a £7,900 exemption so if the ownership of the assets is divided between them, it is possible to claim exemption on gains up to £15,800 jointly in the tax year. Any remaining gain is chargeable as though it were the top slice of the individual's income; therefore according to his or her circumstances it might be charged at 10%, 22% or 40%.

### Self-employed NIC rates (from 6 April 2000)

*Class 2 rate*
Charged at £2.00 per week. If earnings are below £4,615 per annum averaged over the year, ask the DSS about 'small income exception'. Details are in leaflet CA02.

*Class 4 rate*
Business profits up to £4,615 per annum are charged at NIL. Annual profits between £4,615 and £30,940 are charged at 8% of the profit. There is a charge on profits over £30,940. Class 4 contributions are collected by the Inland Revenue along with the income tax due.

*Capital allowances (depreciation) rates*

| | |
|---|---|
| Plant and machinery: | 25% (40% first-year allowance is available for certain small businesses) |
| Business motor cars - cost up to £12,000: | 25% |
| - cost over £12,000: | £3,000 (maximum) |
| Industrial build | 4% |
| Commercial and industrial buildings in Enterprise Zones: | 100% |
| Computers and software equipment | 100% |

THE CONSTRUCTION INDUSTRY TAX DEDUCTION SCHEME

**General**

The new Construction Industry Tax Deduction Scheme is known as the 'CIS' scheme and replaced the old '714' scheme. As the scheme operates whenever a contractor makes a payment to a sub-contractor, the businessman should visit his local income tax enquiry office and obtain copies of the Inland Revenue booklet IR 14/15 (CIS) and leaflet IR 40 which will explain the conditions under which the Inland Revenue will issue a registration card or (CIS6) certificate and precisely when the scheme applies.

Everyone who carries out work in the Construction Industry Scheme must hold a registration card (CIS4) or a tax certificate (CIS6). Certain larger companies use a special certificate (CIS5).

If the sub-contractor has a registration card but does not hold a valid tax certificate (CIS6) issued to him by the Inland Revenue, then the contractor *must* deduct 18% tax from the whole of any payment made to him (excluding the cost of any materials) and to account to the Inland Revenue for all amounts so withheld.

To enable the subcontractor to prove to the Inspector of Taxes that he has suffered this tax deduction, the contractor must complete the three-part tax payment voucher (CIS25) showing the amount withheld. These vouchers must be carefully filed for production to the Inspector after the end of the tax year along with the tax return. Any tax deducted in this way over and above the sub-contractor's agreed liability for the year will be repaid by the Inland Revenue. If he holds a (CIS6) certificate the payment may be made in full without deducting tax.

A small business that does work only for the general public and small commercial concerns is outside the scheme and does not need a certificate to trade. If, however, it engages other contractors to do jobs for it, the business would have to register under the scheme as a contractor and deduct tax from any payment made to a sub-contractor who did not produce a valid (CIS6) certificate. If in doubt, consult your accountant or the Inland Revenue direct.

## VAT

The general rule about liability to register for VAT is given in the VAT office notes. It is possible to give here only a brief outline of how the tax works. The rules that apply to the construction industry are extremely complex and all traders must study *The VAT Guide* and other publications.

Registration for VAT is required if, at the end of any month, the value of taxable supplies in the last 12 months exceeds the annual threshold or if there are reasonable grounds for believing the value of the taxable supplies in the next 30 days will exceed the annual threshold.

Taxable supplies include any zero-rated items. The annual threshold is £55,000. The amount of tax to be paid is the difference between the VAT charged out to customers *(output tax)* and that suffered on payments made to suppliers for goods and services *(input tax)* incurred in making taxable supplies. Unlike income tax there is no distinction in VAT for capital items so that the tax charged on the purchase of, for example, machinery, trucks and office furniture, will normally be reclaimable as *input tax*.

VAT is payable in respect of three monthly periods known as 'tax periods'. You can apply to have the group of tax periods that fits in best with your financial year. The tax must be paid within one month of the end of each tax period. Traders who receive regular repayments of VAT can apply to have them monthly rather than quarterly. Not all types of goods and services are taxed at 17.5% (i.e. the standard rate). Some are exempt and others are zero-rated.

**Zero-rated**

This means that no VAT is chargeable on the goods or services, but a registered trader can reclaim any *input* tax suffered on his purchases. For instance, a builder pays VAT on the materials he buys to provide supplies of constructing but if he is constructing a new dwelling house, this is zero rated. The builder may reclaim this VAT or set it off against any VAT due on standard rated work.

**Exempt**

Supplies that are exempt are less favourably treated than those that are zero rated. Again no VAT is chargeable on the goods or services but the trader cannot reclaim any *input* tax suffered on his purchases.

**Standard-rated**

All work which is not specifically stated to be zero rated or exempt is standard-rated, i.e. VAT is chargeable at the current rate of 17.5% and the trader may deduct any *input* tax suffered when he is making his return to the Customs and Excise. If for any reason a trader makes a supply and fails to charge VAT when he should have done so (e.g. mistakenly assuming the supply to be zero rated), he will have to account for the VAT himself out of the proceeds. If there is any doubt about the VAT position, it is safer to assume the supply is standard rated, charge the appropriate amount of VAT on the invoice and argue about it later.

**Time of supply**

The *time* at which a supply of goods or services is treated as taking place is important and is called the 'tax point'. VAT must be accounted for to the

Customs and Excise at the end of the accounting period in which this 'tax point' occurs. For the supply of goods which are 'built on site', the 'basic tax point' is the date the goods are made available for the customer's use, whilst for *services* it is normally the date when all work except invoicing is completed.

However, if you issue a tax invoice or receive a payment before this 'basic tax point' then that date becomes a tax point. In the case of contracts providing for stage and retention payments, the tax point is either the date the tax invoice is issued or when payment is received, whichever is the earlier.

All the requirements apply to sub-contractors and main contractors and it should be noted that, when a contractor deducts income tax from a payment to a sub-contractor (because he has no valid CIS6) VAT is payable on the full gross amount *before* taking off the income tax.

### Annual accounting

It is possible to account for VAT other than on a specified three month period. Annual accounting provides for nine equal installments to be paid by direct debit with annual return provided with the tenth payment. £300,000.

### Cash accounting

If turnover is below a specified limit, currently £600,000, a taxpayer may account for VAT on the basis of cash paid and received. The main advantages are automatic bad debt relief and a deferral of VAT payment where extended credit is given.

### Bad debts

Relief is available for debts over 6 months.

# Part Four

## GENERAL DATA

## GENERAL CONSTRUCTION DATA

### The metric system

Linear

| | | |
|---|---|---|
| 1 centimetre (cm) | = | 10 millimetres (mm) |
| 1 decimetre (dm) | = | 10 centimetres (cm) |
| 1 metre (m) | = | 10 decimetres (dm) |
| 1 kilometre (km) | = | 1000 metres (m) |

Area

| | | |
|---|---|---|
| 100 sq millimetres | = | 1 sq centimetre |
| 100 sq centimetres | = | 1 sq decimetre |
| 100 sq decimetres | = | 1 sq metre |
| 1000 sq metres | = | 1 hectare |

Capacity

| | | |
|---|---|---|
| 1 millilitre (ml) | = | 1 cubic centimetre (cm3) |
| 1 centilitre (cl) | = | 10 millilitres (ml) |
| 1 decilitre (dl) | = | 10 centilitres (cl) |
| 1 litre (l) | = | 10 decilitres (dl) |

Weight

| | | |
|---|---|---|
| 1 centigram (cg) | = | 10 milligrams (mg) |
| 1 decigram (dg) | = | 10 centigrams (mcg) |
| 1 gram (g) | = | 10 decigrams (dg) |
| 1 decagram (dag) | = | 10 grams (g) |
| 1 hectogram (hg) | = | 10 decagrams (dag) |

### Conversion equivalents (imperial/metric)

Length

| | | |
|---|---|---|
| 1 inch | = | 25.4 mm |
| 1 foot | = | 304.8 mm |
| 1 yard | = | 914.4 mm |
| 1 yard | = | 0.9144 m |
| 1 mile | = | 1609.34 m |

Area

| | | | |
|---|---|---|---|
| | 1 sq inch | = | 645.16 sq mm |
| | 1 sq ft | = | 0.092903 sq m |
| | 1 sq yard | = | 0.8361 sq m |
| | 1 acre | = | 4840 sq yards |
| | 1 acre | = | 2.471 hectares |

Liquid

| | | | |
|---|---|---|---|
| | 1 lb water | = | 0.454 litres |
| | 1 pint | = | 0.568 litres |
| | 1 gallon | = | 4.546 litres |

Horse-power

| | | | |
|---|---|---|---|
| | 1 hp | = | 746 watts |
| | 1 hp | = | 0.746 kW |
| | 1 hp | = | 33,000 ft.lb/min |

Weight

| | | | |
|---|---|---|---|
| | 1 lb | = | 0.4536 kg |
| | 1 cwt | = | 50.8 kg |
| | 1 ton | = | 1016.1 kg |

**Conversion equivalents (metric/imperial)**

Length

| | | | |
|---|---|---|---|
| | 1 mm | = | 0.03937 inches |
| | 1 centimetre | = | 0.3937 inches |
| | 1 metre | = | 1.094 yards |
| | 1 metre | = | 3.282 ft |
| | 1 kilometre | = | 0.621373 miles |

Area

| | | | |
|---|---|---|---|
| | 1 sq millimetre | = | 0.00155 sq in |
| | 1 sq metre | = | 10.764 sq ft |
| | 1 sq metre | = | 1.196 sq yards |
| | 1 acre | = | 4046.86 sq m |
| | 1 hectare | = | 0.404686 acres |

Liquid

| | | | |
|---|---|---|---|
| | 1 litre | = | 2.202 lbs |
| | 1 litre | = | 1.76 pints |
| | 1 litre | = | 0.22 gallons |

Horse-power

| | | |
|---|---|---|
| 1 watt | = | 0.00134 hp |
| 1 kw | = | 134 hp |
| 1 hp | = | 0759 kg m/s |

Weight

| | | |
|---|---|---|
| 1 kg | = | 2.205 lbs |
| 1 kg | = | 0.01968 cwt |
| 1 kg | = | 0.000984 ton |

**Temperature equivalents**

In order to convert Fahrenheit to Celsius deduct 32 and multiply by 5/9. To convert Celsius to Fahrenheit multiply by 9/5 and add 32.

| **Fahrenheit** | **Celsius** |
|---|---|
| 230 | 110.0 |
| 220 | 104.4 |
| 210 | 98.9 |
| 200 | 93.3 |
| 190 | 87.8 |
| 180 | 82.2 |
| 170 | 76.7 |
| 160 | 71.1 |
| 150 | 65.6 |
| 140 | 60.0 |
| 130 | 54.4 |
| 120 | 48.9 |
| 110 | 43.3 |
| 100 | 37.8 |
| 90 | 32.2 |
| 80 | 26.7 |
| 70 | 21.1 |
| 60 | 15.6 |
| 50 | 10.0 |
| 40 | 4.4 |
| 30 | -1.1 |
| 20 | -6.7 |
| 10 | -12.2 |
| 0 | -17.8 |

**Areas and volumes**

| Figure | Area | Perimeter |
|---|---|---|
| Rectangle | Length × breadth | Sum of sides |
| Triangle | Base × half of perpendicular height | Sum of sides |
| Quadrilateral | Sum of areas of contained triangles | Sum of sides |
| Trapezoidal | Sum of areas of contained triangles | Sum of sides |
| Trapezium | Half of sum of parallel sides × perpendicular height | Sum of sides |
| Parallelogram | Base × perpendicular height | Sum of sides |
| Regular polygon | Half sum of sides × half internal diameter | Sum of sides |
| Circle | pi × radius$^2$ | pi × diameter or pi × 2 × radius |

| Figure | Surface area | Volume |
|---|---|---|
| Cylinder | pi × 2 × radius$^2$ × length (curved surface only) | pi × radius$^2$ × length |
| Sphere | pi × diameter$^2$ | Diameter3 × 0.5236 |

**Mild steel cisterns**

| Length mm | Width mm | Depth mm | Capacity mm |
|---|---|---|---|
| 457 | 305 | 305 | 18 |
| 610 | 305 | 371 | 36 |
| 610 | 406 | 371 | 54 |
| 610 | 432 | 432 | 68 |
| 610 | 457 | 482 | 86 |
| 686 | 508 | 508 | 114 |
| 736 | 559 | 559 | 159 |
| 762 | 584 | 610 | 191 |
| 914 | 610 | 584 | 227 |
| 914 | 660 | 610 | 264 |

**Plastic cisterns**

| Ref. | Capacity litres | Capacity gallons | Weight kg |
|---|---|---|---|
| PC4 | 18 | 4 | 0.85 |
| PC15 | 68 | 15 | 2.95 |
| PC25 | 114 | 25 | 3.40 |
| PC40 | 182 | 40 | 6.35 |

**Roof drainage**

| | Area m2 | Pipe mm | Gutter mm |
|---|---|---|---|
| One end outlet | 15 | 50 | 75 |
| | 38 | 68 | 100 |
| | 100 | 110 | 150 |
| Centre outlet | 30 | 50 | 75 |
| | 75 | 68 | 100 |
| | 200 | 110 | 150 |

# Index


Accountant, 207-208
Alterations and repairs, 73-81
Aluminium rainwater goods
  gutters, 23
  pipes, 19-20
Appliance valves, 63

Banks, 205, 215-216
Baths, 9-10
Bath accessories, 10-11
Bath panels, 10
Bathroom projects, 125-162
Basins, 11-12
Bidets, 13
Boilers
  gas, 66-67
  oil, 68
Break-even point, 223-224
Business links, 204
Business matters, 203-234
Business plan, 209-214

Capital Gains Tax, 230
Cashflow, 220
Cast iron
  drains, 33-39
  gutter, 24-25
  rainwater pipe, 21-22
Central heating projects, 175-193
Cisterns, 64
Chases, 18
Check valves, 62-63
Cold water storage tanks, 64
Construction Industry Tax
  Deduction Scheme, 231-232
Cylinders, 65

Debtor days, 221
Department of Transport and
  the Regions, 207
Department of Trade
  and Industry, 207
Drainage
  above ground, 26-32
  below ground, 33-39

External waste systems, 163-173

Finance
  banks, 215-216
  Business Start-up Scheme, 217
  Enterprise Investment
    Scheme, 216
  hire purchase/leasing, 216
  personal funds, 214-215
  Royal Jubilee and Princes
    Trust, 217
  venture capital, 216
Fixed Profit Car Scheme, 228

Gas boilers, 66-67
Gate valves, 50, 61-62
General data, 237-242
Gullies, 38-39
Gutters, 22-24

Health and Safety
  Executive, 209
Holes, 16-17
Holes and chases, 16-18
Hot and cold water projects,
  195-200
Hot and cold water services,
  40-65
Hot water cylinders, 65

Inspector of Taxes, 205
Insulation
  jacket, 65
  pipe lagging, 65
Insurance broker, 208-209

Labour, xvii
Leadwork, 3-8
  aprons, 5
  cappings, 6
  dormers, 3, 4
  dots, 8
  dressings, 8
  edges, 7-8
  flashings, 4-5
  flat roof, 3
  gutters, 7
  hips, 6
  kerbs, 6-7
  ridges, 7
  sills, 5-6
  sheet coverings, 3-8
  slates, 8
  sloping roof, 3
  soakers, 8
  valleys, 7
Limited company, 225
Local authorities, 206-207

Marketing, 219
Materials, xvi-xvii

National Insurance
  contributions, 230-231

Oil boilers, 68
Oil storage tanks, 68-69
Overflows, 31
Outlets, 25
Overheads and profit, xvii
Over-trading, 222-223

Partnership, 225
Pipes
  rainwater, 19-22
  soil, 28-31
  waste, 26-28
Plug cocks, 51
Premises, 228
Profitability, 221-222
Project costs
  bathrooms, 125-162
  central heating, 175-193
  external waste, 163-173
  hot and cold water, 195-200
PVC-U
  gutters, 22
  overflows, 31
  pipes, 19
  soil pipes, 28-29

Radiators, 70-72
Radiator valves, 51-52
Rainwater goods, 19-25
Rainwater gutters
  aluminium, 23
  cast iron, 24
  PVC-U, 22
Rainwater pipes
  aluminium, 19-20
  cast iron, 21-22
  PVC-U, 19
Rainwater outlets, 25
Ridges, 7

Sanitary fittings, 9-15

Self-assessment, 226
Servicing valves, 63
Showers, 15
Shower valves, 15
Sloping roof, 3
Sills, 5-6
Sinks, 12-13
Slates, 8
Soakers, 8
Soil pipes
  cast iron, 29-31
  PVC-U, 28-29
Sole trader, 225
Solicitor, 208
Stock turn, 221
Stop valves, 49-50, 60-61
Storage tanks, 64

Tanks
  cold water, 64
  oil storage, 68-69
Taps, 14-15
Taxation
  allowances, 229
  business, 227-228
  rates, 229-230

**Taxation (cont'd)**

  correspondence, 226-227
  dates due, 227
  entertainment, 229
  Training and Enterprise Councils, 204
Traps, 31-32

Unit rates, 1-81
Urinals, 14

Valleys, 7
Valves
  appliance, 63
  check, 62-63
  gate, 50, 61-62
  radiator, 51-52
  servicing, 63
  shower, 15
  stop, 49-50, 60-61
VAT, 206, 232-234
Vehicles, 228-229

Wash basins, 11-12
WCs, 13
Waste pipes, 26-31